Association of Reptilian and Amphibian Veterinarians Conference 2016 (ARAV 2016)

Held at ExoticsCon 2016

Portland, Oregon, USA
27 August – 1 September 2016

ISBN: 978-1-5108-4289-2

Printed from e-media with permission by:

Curran Associates, Inc.
57 Morehouse Lane
Red Hook, NY 12571

Some format issues inherent in the e-media version may also appear in this print version.

Printed by Curran Associates, Inc. (2017)

For permission requests, please contact Association of Reptilian & Amphibian Veterinarians at the address below.

Association of Reptilian & Amphibian Veterinarians
5003 SW 41st Blvd
Gainesville FL 32608-4930
USA

Phone: 480-703-4941

info@arav.org

Additional copies of this publication are available from:

Curran Associates, Inc.
57 Morehouse Lane
Red Hook, NY 12571 USA
Phone: 845-758-0400
Fax: 845-758-2633
Email: curran@proceedings.com
Web: www.proceedings.com

TABLE OF CONTENTS

SECTION 1: TECHNICIAN SESSIONS

SECTION 2: PRACTICE MANAGEMENT SESSIONS

SECTION 24: ARAV PLENARY SESSIONS

SECTION 25: ARAV INFECTIOUS DISEASES

SECTION 26: ARAV ANESTHESIA AND ANALGESIA

SECTION 27: ARAV MASTER CLASSES AND ROUNDTABLES

SECTION 28: ARAV CLINICAL REVIEWS AND CARE IN CAPTIVITY

SECTION 29: ARAV THERAPEUTICS AND CASE STUDIES

SECTION 30: ARAV CONSERVATION

SECTION 31: ARAV POSTERS

SECTION 32: SATURDAY HYBRID MASTER CLASSES

SECTION 33: WEDNESDAY HYBRID TRACK

SECTION 34: HYBRID POSTERS

Section 1

Technician Sessions

A Chicken Just Walked In Through the Front Door, Now What?

Janet L. Jones, LVMT

Session #080

Affiliation: From the Veterinary Medical Center, University of Tennessee, Knoxville, TN 37996, USA.

Abstract: Chickens have become a popular backyard pet. Many cities have now allowed people to keep chickens in their backyard. It has become more common for the veterinary practice to have a chicken present for illness or emergency. A chicken is not triaged any differently than any other domestic animal. It is important to understand what is normal, including behavior, before being able to identify what is abnormal. The history is the most important starting point. The physical exam starts with visual observation, then with gentle restraint, start with the head and work your way to the back end. Treatments will vary depending on if the bird is for egg production, meat production, show, or as a pet only.

When that client walks into your clinic with a sick chicken, that sick bird should be triaged just like any other domestic animal. Owners are often very attached to their pet chickens so emotions can run high with some clients. The most important part of the examination is obtaining a thorough history. The next step is to do a complete physical examination.[1] Most chickens will tolerate minimal physical restraint. By starting with the head and working down to the tail, you should be able to examine all the important body parts of the bird.

The most important part of the physical examination is to obtain a complete history. Questions to be asked include the signalment of the bird and determination of its use. It is important to know how many are in the flock, how long they have been in the flock together, where they were obtained, and if any vaccinations were given. Housing questions include size and composition, space of the coop and range, substrate and perch size, are they brought in at night? Questions to ask about nutrition include type, brand, amount, where purchased, how old is the food, any treats or supplements, and any medicated foods? If the patient is a laying hen, questions to be asked should include how many boxes and sizes, and the hens laying history? When addressing the presenting complaint, ask how long it has been occurring, any other birds are affected, and if the condition has progressed?

Before restraining the bird, a visual examination should be performed from a distance. Note the mentation of the bird. Chickens will try to act normal even when they are very ill. They put on this show so as not to get picked on by other birds in the flock or get picked off by a predator. A healthy bird will be curious and alert of their surroundings. A sick bird will hold its head down, eyes partially closed, and not standing upright. Observe in the cage the droppings for consistency, blood, or abnormal size or smell. There are three components to the droppings; the clear urine portion, the dark colored stool portion and the white urate portion. Chicken restraint starts with controlling their wings. Gently hold their wings to the body in a natural folded position. Do not place heavy restraint across the keel as chicken do not have a diaphragm. Usually the least amount of restraint possible is best for the birds comfort. Place a towel down on the exam table to prevent foot slippage. Examine from head to tail and leave the oral exam last as most chickens do not like having an oral exam and will get stressed during this part of the examination.

Examine the head for symmetry of the beak, eyes, sinuses and nostrils while looking for discharge, crusts, scratches, scabs, swellings, and discolorations. The beak should come to a point. The iris should both be the same color. Examine the ears by gently moving away the feathers and look for any discharge, blood, or parasites. The comb should be firm and red. A capillary refill time can be performed on the comb with a refill time of 2 seconds or less. The comb should not be pale or flopped over to one side, not unless it is normal for the bird. Hydration status can be obtained from the basilic vein be pressing on the vein and noticing refill time. A hydrated bird will refill instantly, in a dehydrated bird you will be able to see the vein refill and if the bird is in shock or over 10% dehydrated there is a one second refill time. Palpate the crop; it should feel soft and fluctuant. Birds lack organized lymph nodes so there should be no subcutaneous masses. The keel should be straight and you should be able to feel the pectoral muscles on either side.

Auscult the heart and listen to heart sounds by placing the stethoscope on either side of the keel. Birds have a four chambered heart and their rate tends to be faster than most mammals of about 150-250 beats per minute. You should not hear any lung sounds in birds and their respiratory rate is usually 15-30 breaths per minute. Temperature is typically not taken in birds. The feathers should lay down smooth. Lift them up and check for ecto parasites. All joints should be examined and palpated for range of motion, stiffness, crackles, and clicks. The coelomic cavity should be palpated and should be soft and doughy. It should not be extended or fluid filled. The vent should be clean and dry. The bird should be lifted up to expose the plantar surface of the feet looking for scabs, callouses, swellings, dark areas, cuts, or keratin overgrowth.

The last part of the exam is the oral exam. Gently open the bird's mouth to evaluate for white plaques, masses, abnormal smell, and evaluation of the choanal slit and papillae. Accurately weigh the bird on a gram scale to calculate dosages.

When you come up with your treatment plan, keep in mind that there are certain drugs that are illegal to use in chickens. You can access all of these drugs from Food Animal Residue Avoidance Databank (FARAD) website at www.farad.org.[2] They have a databank listing the withdrawal times for drugs used in food-producing animals in the United States.

References

1. Greenacre CB, Morishita TY. *Backyard Poultry Medicine and Surgery: A Guide for the Veterinary Practitioner*. Ames, IA: John Wiley & Sons; 2015.

2. Food Animal Residue Avoidance Databank (FARAD). Available at: http://www.farad.org. Accessed May 24, 2016.

Chicken Breeds: Who Lays The Best Eggs?

Janet L. Jones, LVMT

Session #081

Affiliation: From the Veterinary Medical Center, University of Tennessee, Knoxville, TN 37996, USA.

Abstract: Chicken breeds can be categorized as egg, meat, dual purpose, and ornamental breeds. All of today's many breeds of chickens came from a single origin, the red jungle fowl. The most productive egg layers are the hybrids. Some hybrid pullets start laying when they are 17 weeks of age, but most breeds take more than 26 weeks before they mature and start laying. Egg coloration ranges from white, brown, to greenish or bluish eggs. Other considerations when choosing the right chicken is temperament of the hen and temperature tolerance.

Chickens have coexisted with people for centuries providing both meat and eggs. Today's chickens are now also used as a companion animal. All of today's breeds of chickens came from the red jungle fowl of Southeast Asia. Today's breeds of chickens can be divided into egg breeds, meat breeds, dual purpose breeds, and ornamental breeds. A survey conducted in 2010 found that 47% of chicken owners have them for eggs; whereas, less than 10% sell them for meat.

The most productive egg layers are hybrids, including the Hy-Line Brown, California White, Golden Comet, Cherry Egger, and the Indian River. The Brown sex Links lay zmostly extra-large eggs. Some hybrid pullets can start laying when they are only 17 weeks old. Other breeds take more than 26 weeks to mature and start laying eggs. Some of the breeds that lay the darkest brown eggs are the Marans, Barnevelders and Welsummers. The Ameraucanas and Aracanus lay greenish or bluish eggs. In most cases, chickens with red ears lay brown eggs and hens with white ears lay white eggs. Chickens that are allowed to range on pasture produce the most nutritious and flavorful eggs.

Some specific traits of the more common egg breeds are the Ameraucanas are the Easter egg layers. They lay about 3 medium green to blue colored eggs per week. They have a calm and gentle disposition and are a very hardy bird. The Hamburg lays about 4 small to medium white eggs per week and tend to be a bit more flighty and love to free roam. The Lakenvelder lays about 3 medium white eggs per week and the hens are non-broody but can be good flyers. The leghorn is the best of the layers as they lay about 5 white extra-large eggs per week. They are a very docile breed that handle confinement well. The Maran produces 3 large dark brown eggs per week and the hens have a calm temperament and adapt to confinement well. The Sicilian Buttercups lay about 2 small white eggs per week and the hens are very active birds.

Other considerations when choosing the right chickens for a backyard flock is temperature tolerance. The breeds that tolerate cold weather the best are Chanteders, Buckeyes, Brahmas and Javas. Where heat is more of a problem, those hens that tolerate hot environments are Jungle Fowl, Malays, Sumatras, Javas and Cubalayas. The breeds that tend to be the most broody are the Silkies, Aseels, Modern Games, Old English Games and Cochins. The tendency to be broody with chickens has been bred out of many breeds because hens stop laying eggs for a period of time when they are broody even if they are not hatching eggs.

There are many different factors to consider when keeping chickens for egg consumption and there are many different breeds as well.[1-2] By taking into account the environment, size of enclosure, and if chickens are wanted as pets as well as eggs production; you now have the information available to help make the decision as to what is the best chicken for desired traits.

References

1. Greenacre CB, Morishita TY. *Backyard Poultry Medicine and Surgery: A Guide for the Veterinary Practitioner*. Ames, IA: John Wiley & Sons; 2015.

2. Griepentrog, T. Best chicken breeds for backyard flocks: our latest survey results can help you choose the best chickens for eggs, meat, temperament and more. *Mother Earth News.* April-May 2010: 30.

Chicken Diseases: What the Backyard Chicken Owner Needs To Know

Janet L. Jones, LVMT

Session #083

Affiliation: From the Veterinary Medical Center, University of Tennessee, Knoxville, TN 37996, USA.

Abstract: When chickens present at the clinic they are often showing general signs of disease. A compete and through history are the most important first step in diagnosis of the patient. Respiratory diseases are the most common diseases of poultry with non-specific respiratory disease being extremely common. Other very common presentations are trauma and husbandry issues. Other conditions range from musculoskeletal, dermatologic, gastrointestinal, reproductive and zoonotic diseases.

Backyard poultry present to the clinic when the patient has not been doing well and often for a number of days. General signs of illness with backyard poultry are usually decreased appetite, lethargic, distance from the rest of the flock, not perching, and sitting with the head down and eyes closed. The most important part of the initial examination is getting an accurate, compete, and through history from the owner.

The most common respiratory disease of chickens is non-specific respiratory disease. This usually occurs in the spring and fall of the year when the weather changes. The patient presents with tracheal rales, snicks, sneezes, and coughing. They continue to be alert, active, and continue to eat and drink. If left untreated, the respiratory signs usually subside in 7-10 days. Mycoplasmosis is also very common in chickens. They present with a cough and with secondary E. coli infection, they may also have decreased activity, look "puffed up", and appear sleepy and reluctant to move. Clinical signs can reoccur after discontinuation of antibiotics. With infectious coryza there is acute death or the development of respiratory signs including sick birds, oculonasal discharge, facial edema, and/or infraorbital sinuses that are swollen. The last common respiratory disease is fowl cholera. These birds present with subcutaneous masses, swollen eyes, ears, or wattles. Depopulation is usually recommended to remove disease carriers.

Trauma is a very common presenting complaint usually as a result of predator attack. It is important during your physical examination that while you are palpating the patient that you part the feathers and check the skin carefully. Chickens have very thin skin and small lesions and puncture wounds can be very easily overlooked. Fly strike is common in debilitated birds with open wounds or fecal accumulation near the vent. Those birds that are severely ill need to be stabilized and correction of shock before extensive wound management under sedation or anesthesia is to be performed. For husbandry presentations, usually pododermatitis due to moisture in the bedding is a common problem as well as those birds that are kept on concrete for long periods of time. High levels of ammonia and dust can also cause non-specific respiratory disease.

For musculoskeletal diseases, when husbandry has been ruled out for pododermatitis, septic joint can be a common presentation for lameness in a chicken. This is usually the result of bacteria in a joint due to trauma, diarrhea, or respiratory disease that allowed bacteria access to the joint. These birds present with lameness either in one or multiple joints, and an enlarged, warm joint. Prognosis can be fair to poor and long term therapy can result in arthrodesis and limited range of motion. Merek's disease affects only chickens and these birds present with either unilateral or paralysis with one leg stretched forward and the other leg pointing back. These birds are typically between 10-20 weeks of age. The cause of this disease is a herpesvirus and once a bird is infected they shed the virus indefinitely. Hens with reproductive diseases have a distended coelom due to fluid or soft tissue/

organ enlargement. They can present with lameness resulting from weakness, illness, and simply as a result of an enlarged coelomic cavity that prevents normal ambulation. Valgus deformities result from lateral deviation of the intertarsal joint, tibiotarsus, the hip, or a combination of these. Treatment can involve slowing down the growth of the bird, or in the case of splay leg by bandaging the leg in its normal position. Slipped tendon is luxation of the gastrocnemius tendon forcing the leg into a vulgus position with a swollen hock. Surgery can be performed in young birds if caught early enough.

For dermatologic diseases in chickens, avian pox presents as round "pox-like" lesions of the integument of the mostly unfeathered region of the head or neck. Lesions can also be seen on the feet or vent. The patient may also have decreased weight gain and egg production. Affected birds should be isolated. Lice signs include hyperemia and irritation of the skin with small scabs and clots. Feathers can look moth-eaten and nits are laid in clumps at the base of the feathers along the ventrum. Mites are much smaller than lice, and feed on blood, feathers, and skin. Appearance is variable. Treatment for both lice and mites is usually ivermectin.

The most common gastrointestinal disease of hens is either a pendulous crop which the cause is unknown or the patient presents with an enlarged and fluid filled crop. Crop impaction can occur when chickens are exposed to new environmental substrates, such as grass, newspaper, sawdust shaving/wood chips, and feathers. Surgical removal is usually necessary. Coccidiosis is a common intestinal parasite of chickens and they present with pale combs and wattles from blood loss in the intestines, ruffled feathers, depression, blood in droppings, and shivering. Mortality rate is high in young birds. Husbandry can play a role in control of this parasite along with anticoccidial drugs.

For zoonotic diseases, *Salmonella* needs to be addressed with owners of backyard chickens. Most of the estimated 1.4 million annual cases of human salmonellosis in the United States result from ingestion of contaminated food including eggs. It is transmitted via the fecal-oral route. *Salmonella* naturally lives in the intestine of poultry and is considered normal flora in healthy, live poultry.

This list of diseases is the most common that are typically seen in practice.[1,2] There is a wide array of both common and uncommon diseases of chickens. There are a number of resources in both book format and online resources to help guide you through the diagnosis pathway. The University of Florida has a viewable table that can be helpful in determining diseases based on clinical signs.

References

1. Greenacre CB, Morishita TY. *Backyard Poultry Medicine and Surgery: A Guide for the Veterinary Practitioner*. Ames, IA: John Wiley & Sons; 2015.

2. University of Florida, IFAS Extension. *Common Poultry Diseases.* Available at: http://edis.ifas.ufl.edu. Accessed May 24, 2016.

Chicken Husbandry: What Owners Need To Know About Housing and Predators

Janet L. Jones, LVMT

Session #084

Affiliation: From the Veterinary Medical Center, University of Tennessee, Knoxville, TN 37996, USA.

Abstract: There are many considerations when backyard chickens are kept as pets. Owners must be aware of the time, effort, and knowledge required taking care of their chickens. Not only must the coop provide adequate space and ventilation, it must also be predator proof as just about any predator will take poultry if they get a chance. On the list of predators are dogs, they tend to kill birds as a result of play with the chicken and are typically not maimed as in a predator attack.

Basic management of chickens includes housing, chick purchasing, brooding, daily care, and predator options.[1-2] Good husbandry and housing is as vital for the wellbeing of backyard chickens as it is for other livestock. When choosing a breed of chicken, the outside environment must be taken into consideration. Most chickens prefer to have free range during the day and secure housing to roost at night.

Housing design will affect the chickens care, comfort, welfare, and well-being. Housing should provide shelter from the wind and rain, protection against predators when the birds are shut in, a dry and safe area to sleep, and a secluded area where hens can lay eggs. The coop can range from a simple design of a small plastic pod to an elaborate house. The ideal structure would have a cement floor and be insulated with washable walls. Depending of the size of the hens, usually a 12” × 12” laying box is sufficient for most hens. Any openings in the chicken coop should be covered with hard wire cloth and the opening door should have a screen door. This may not always be possible with an outdoor run and the risks must be weighed.

Ventilation is important for moisture removal, excess heat removal, exchange of gases produced by the litter, and providing fresh air. The winter months can be the most difficult to maintain ventilation and moisture control. Usually with air inlets along the roof line and an opening where air can come in will achieve this. The temperature of the coop should be between 50° and 70°F. Birds can get frost bite with their heads and feet being the most common to get frostbite. Birds that get heat stress will pant, spread their wings, increase water consumption, and decrease food consumption. Most chickens can grow sufficiently with normal daylight and do not need any special lighting. Birds that are egg producers should have a day length of 15 hours and 8 hours of dark. Egg production will naturally increase in the spring and decrease in the fall and winter months. The ideal bedding material should be absorbent, loose, and fairly inexpensive. The most common used is pine wood shavings. Hard wood shavings are not recommended due to potential presence of molds and fungus which can result in respiratory infections.

When purchasing chicks, obtaining them from a reputable hatchery is recommended along with paying the extra few dollars and having them vaccinated for Merek’s disease. The brooder pen should be set up 48-72 hours prior to chick arrival. The heat bulb should be centrally located with feed and water alternating around the heat source. Start with a 95°F temperature for the first week and decrease by 5°F each week until outdoor temperature is met. The chicks should be evenly distributed throughout the brooder. They should be fed a commercially made medicated starter diet. When they are in their grow-out stage at about 6 weeks of age, they can be given limited outdoor access as they will be predominantly feathered and can tolerate the environment better. The birds will learn to enter the coop at night as sunset approaches. At adulthood, usually 16-24 weeks of age, the hen will start laying. No additional management changes need to be made. Birds establish a social hierarchy and disrupting this can cause aggressive fighting between birds. Hens may stop laying eggs while going through a molt which can last up to 6 weeks. They should be fed a commercially made laying formula.

As a general rule, the space requirements for most birds are 377 square inches per adult bird. If chickens are too densely populated, they can become cannibalistic and peck at each other if they do not have enough feed, water, nest boxes, incorrect lighting, abrasions or tears as a result of injury or mating, dietary deficiencies, prolapse, or meanness of a breed. Usually increasing the space and additional enrichment may help reduce these behaviors. A broody hen will stop laying eggs, sit on a nest, and become more aggressive as she protects her eggs. It is recommended to remove eggs on a daily basis. Egg eating can develop from overcrowding, uneven nest space, nutritional deficiency, too bright of light, or disposing of broken or cracked eggs in the coop.

For predator control, ensure that all holes or other access into the chicken coop is secured. Raccoons, opossums, mink, skunks, foxes, coyotes, or weasels may find ways to enter the chicken coop and pull chickens through wire and holes to consume them. Wire fences should be buried 8 inches deep to ensure that predators are unable to dig under the fence. Bushes and shrubs can provide access to aerial attacks by raptors. Dogs can be responsible for what appears to be a predator attack but is the result of trying to play with the chickens.

References

1. Greenacre CB, Morishita TY. *Backyard Poultry Medicine and Surgery: A Guide for the Veterinary Practitioner*. Ames, IA: John Wiley & Sons; 2015.

2. Wallace-Houghton, J., Lister, S. Backyard poultry husbandry and general management. *In Practice*. 2012;34:136-145.

Chicken Anesthesia: Are Chickens Different Than Other Birds?

Janet L. Jones, LVMT

Session #082

Affiliation: From the Veterinary Medical Center, University of Tennessee, Knoxville, TN 37996, USA.

Abstract: There may be many circumstances as to why a chicken may need to be under anesthesia. Some of the most common reasons may be for obtaining diagnostic samples, or for medical and surgical treatments. The same basic principles of anesthetic management apply to the chicken patient as well as any other avian patients with small differences in restraint of the poultry patient. Use of monitoring equipment and anesthetic depth are important components to the anesthetic procedure.

Monitoring the depth of anesthesia for your poultry patient can be one of the most challenging tasks during the procedure. There are many different monitors that should be used during the anesthetic procedure as well as using your own observations. The challenge is to keep the patient at a surgical level of anesthesia. By checking the pinch reflex as well as monitoring trends of decreasing or increasing respiratory and cardiac trends, will help keep your poultry patient at that preferred anesthetic plain.

Perioperative considerations include evaluation of all aspects of husbandry including the source and length of ownership, diet and dietary preferences, and previous or current illnesses and therapies. A thorough physical examination is critical in assessing the chickens' physical condition, their cardiopulmonary status, and the severity of illness. Minimum diagnostics should include a packed cell volume (PCV), total solids, and blood glucose. Preferred diagnostics would be a complete white blood cell count (CBC) and a biochemistry. Another preanesthetic consideration is to fast the chicken. Generally a 4-6 hour fast is adequate for crop empting. The crop should be gently palpated prior to an anesthetic event and the head should be elevated during the surgical procedure.

To restrain the chicken for physical examination, generally the least amount of restraint with chickens is the best. By gently placing hands around the chicken's wings and allowing the chicken to stand on an examination table on a towel will eliminate foot slippage and most chickens should stand comfortably. Examination of the oral cavity should be performed last as most chickens do not like having their mouth manually opened. The bird can be picked up by placing the hands around the wings and gently lift off the table to examine the feet. Any premedications should be administered at this time. Once the bird is sedated, wrap a towel around the body to prevent the wings from flapping and using gas anesthesia is preferred. Once the bird is anesthetized, it should be placed on a heated surface where the procedure is to be performed and should not have a righting reflex or flap its wings. The bird should not respond to a firm toe pinch. The respiratory rate and effort must be visualized during the whole anesthetic procedure.[1-3]

Once the bird is laid on the table it is preferred that an intravenous catheter be placed in either the basilic vein or the medial metatarsal vein. An intraosseous catheter can also be placed in either the distal ulna or the proximal tibotarsus. The bird should be intubated using a non-cuffed, appropriately sized endotracheal tube. Using a heated table or additional heat source to place the bird on will help prevent the patient from becoming hypothermic. Once laid in recumbency, the monitoring equipment should be placed on the bird. The most important piece of equipment is the Doppler. The crystal can be placed in the mouth along the palatine artery, on the basilic vessel on the wing, or on the medial metatarsal vessel on the leg. A hand held pulse oximeter should be placed on either the toe or the beak while the rest of the monitoring equipment is being placed on the bird. The use of a multifunction monitor is very beneficial as monitoring ECG, respiratory rate with end–tidal CO2, and pulse oximetry are recommended. Fluids should be administered during the anesthetic period and emergency drugs should be readily available in case of an emergency.

During the anesthetic procedure the respiratory rate and effort must be visually monitored at all times. Birds will usually go into respiratory distress before they go into cardiac distress. If a decrease in respiratory or cardiac rate is appreciated, the anesthetic gas should be adjusted, usually in 0.25% increments. Any avian patient can get cold during an anesthetic event and will need to have the anesthesia turned down during the procedure. Listening to the Doppler is one of the most valuable monitoring methods as any change in cardiac rate or rhythm will be immediately heard. Blood pressure can be monitored during the procedure by placing the appropriate sized cuff on the distal humorous or dorsal metatarsal artery.

Once the anesthetic procedure is completed, the gas should be turned off and the patient left on oxygen. Chickens will feel more comfortable upon recovery if they are placed in sternal and once they start to shake their head, the endotracheal tube should be removed. The endotracheal tube should be examined for any signs of blood. Once the chicken is able to stand and ambulate well on its own, the patient can be place in its cage with food and water. Post-operative analgesics must be considered and appropriate for the surgical procedure performed.

References

1. Gunkel C, Lafortune M. Current techniques in avian anesthesia. *Semin Avian Exot Pet Med.* 2005;14: 263-276.

2. Nevarez JG. Monitoring during avian and exotic pet anesthesia. *Semin Avian Exot Pet Med.* 2005;14: 277-283.

3. Raftery A. Avian anaesthesia. *In Practice*. 2013;35:272-278.

Complete Blood Counts and Normal Hematologic Values in Young Bobwhite Quail

Ian Kanda, RVT,
Jessica Robertson, DVM,
James Meinkoth, DVM, MS, PhD, Dipl ACVP,
Joao Brandao, LMV, MS

Session #056

Affiliation: From the Department of Veterinary Clinical Sciences (Kanda, Robertson, Brandao) and the Department of Veterinary Pathobiology (Meinkoth), Center for Veterinary Health Sciences, Oklahoma State University, 2065 W Farm Road, Stillwater, OK 74078, USA.

The Northern bobwhite quail *(Colinus virginianus*) is a game bird listed as near threatened by the International Union for Conservation of Nature and it is experiencing decline throughout much of its range.[1-5] Species and age-specific hematologic data for bobwhite quail is limited.[6] The purpose of this study is to report hematologic values in healthy young Northern bobwhite quail, using different white blood cell count methodologies.[7,8] In 18 49-day-old captive raised quail, complete blood cell counts were performed as part of a health assessment prior to release into the wild. Data were reviewed and analyzed for the purpose of this study. Values for packed cell volumes (0.36 [0.25-0.45] L/L), total solids (36 [23-41] g/L), mean corpuscular volume (240 [166-327] fL), red and white blood cell counts (1.56 [0.93-2.04] x 10^{12}/L, and 13.8 [3.05-22.0] x 10^{9}/L, respectively) were obtained. A dilution with Natt-Herrick's stain in a manual hemocytometer was used to obtain the red blood cell count. Cell differentials were recorded for each animal. With considerable variability, bobwhite quail of this age were found to be primarily lymphocytic (57 [23.8-81.8] %). White blood cell counts were obtained from blood smears, estimating on both 400x (16.2 [9.2-23.4] x 10^{9}/L) and 1000x (10.8 [4.4-18.4] x 10^{9}/L) magnification. Counts were also made using a manual hemocytometer with blood diluted in phloxine B stain (13.8 [3.05-22.0] x 10^{9}/L). Findings suggest that 1000x magnification used for white blood cell counts underestimates results and 400x magnification overestimates results when compared to the current gold standard utilizing Phloxine B stain.

Acknowledgments: The costs associated with this project were supported by the Kristie Plunkett Wildlife Fund.

References

1. International Union for Conservation of Nature. *Colinus virginianus. The IUCN Red List of Threatened Species*. Available at: http://www.iucnredlist.org/details/22728956/0. Accessed July 3, 2016.

2. Brennan LA. How can we reverse the northern bobwhite population decline? *Wildl Soc Bull.* 1991:544-555.

3. Lusk JJ, Guthery FS, DeMaso SJ. Northern bobwhite (*Colinus virginianus*) abundance in relation to yearly weather and long-term climate patterns. *Ecol Modell.* 2001;146:3-15.

4. Williams CK, Guthery FS, Applegate RD, et al. The northern bobwhite decline: scaling our management for the twenty-first century. *Wildl Soc Bull.* 2004;32:861-869.

5. Williams CK, Lutz RS, Applegate RD, et al. Habitat use and survival of northern bobwhite (*Colinus virginianus*) in cropland and rangeland ecosystems during the hunting season. *Can J Zool.* 2000;78:1562-1566.

6. Ernst R, Coleman T, Kulenkamp A, et al. The packed cell volume and differential leucocyte count of bob-white quail (*Colinus virginianus*). *Poult Sci*. 1971;50:389-392.

7. Campbell TW, Ellis CK. *Avian and Exotic Animal Hematology and Cytology*. 3rd ed. Ames, IA: Blackwell Publishing;2007.

8. Dein FJ, Wilson A, Fischer D, et al. Avian leucocyte counting using the hemocytometer. *J Zoo Wildl Med*. 1994:432-437.

Section 2

Practice Management Sessions

Creating Great Content for Social Media

Laurie Hess, DVM, Dipl ABVP (Avian)

Session #344

Affiliation: From Veterinary Center for Birds & Exotics, Bedford Hill, NY 10507, USA.

First, we have to come to an understanding. Everything we do in the digital space is done to advance your practice. Whether your goals are to grow financially, to treat more pets, to increase adoptions, or to create a more informed community of pet owners, we must proceed with a purpose. That purpose is to have your story heard. This talk will help you understand the factors that you can control (and some you can't) when you consider sharing your message.

Keyword Roots

For years and years, you have heard talk of keywords. In the early days of the Internet, websites could simply drop keywords into a designated section of the website's code appropriately named keywords. This practice helped the web spiders that crawled and inspected websites so that they could be categorized for searches by knowing which sites contained what information. This practice of simply listing keywords was short lived, in that once websites all began listing the same keywords, search engines needed to rely on different indicators. For example, if every veterinary practice listed veterinarian in their keywords, the search engine would return about 30,000 results for practices. How would these be sorted?

Although the game of how keywords are used has changed, they are still at the root of any good search engine optimization (SEO) campaign. I know - you thought this talk was going to be about social media. As I mentioned earlier, every social media effort requires a focus on SEO, for it is about sharing your story with as many current and potential customers as possible. If no one is listening, how valuable is your message?

Know Your Keywords

It is very important that thorough research be conducted to determine what words are being used to find practices like yours. This should be done by a professional firm that has the ability to scour the Internet for information about terms and phrases that are being used successfully in both your campaigns and those of your competitors and to determine the actual phrases used in searches that are leading pet owners in your area to practices like yours. A reliable firm should be able to tell you how much traffic is being generated for specific websites. With this information in hand, you should be able to create a list of desirable keywords. Here are some examples of keyword categories that may work for your practice:

- Practice name (including shortened varieties)
- Best veterinarian (insert town, county, or other geographic identifiers)
- New puppy/kitten/bird/bunny, or other care (insert town, county, or other geographic identifiers)
- Veterinary (insert town, county, or other geographic identifiers)

Do you see a theme developing? Yep...it's all about local search. Google and the other search engines compete to return the best results to the web user's search query. Better results for Google means that a site has greater credibility; greater credibility leads to more frequent usage, which ultimately leads to greater ad sales opportunities for Google. By the way, I refer to Google so frequently and use that term generically to represent all search engines, because most searches (some studies say over 85%) are done using the almighty Google. If Google's goal is to return great information, and they can tell that the searcher is located in a particular geographic region, doesn't it make sense that they would want to respond with businesses that are accessible to the user? You betcha. So, why not make it easy on Google? Tell Google where you are located.

Key Phrases

Research has shown that most searches involve more than one word. Think about your own behavior. If you were looking online for a company to install a privacy fence around your yard, would you search fence? If you did, you would wind up with many undesirable results. Instead, if you are like most, you would be more detailed in your request. Something like: privacy fence companies in Chicago or best privacy fence company in Chicago would work better. Consider this fact when developing your strategy.

Keyword and Key Phrase List

Once you have worked with a professional firm to develop the list of valuable keywords, you will have a system to guide your practice's online content creation efforts. Let's keep this list handy, but put it off to the side while we look at developing a strategy.

Posting Calendar

One of the biggest problems that practices have is knowing what to post when. In many cases, posting is a reaction to the realization that the practice Facebook, blog, or Twitter accounts have been inactive. Inevitably, the person in charge of posting on the practice's social media sites sits in front of the computer to create activity. This type of content creation often results in very low interaction.

Posts should be written weeks in advance and scheduled on a calendar, so that the daily activity involves copying very well developed content from a spreadsheet or word document onto the practice's social media sites. This activity prevents underdeveloped posts but allows for the timely posting of great activities that happen in the practice on a daily basis.

Right about now, a hand goes up in the audience with a question regarding how often a practice should post. My answer is: It depends. I bet you were looking for a more detailed response. The truth is that there is no specific recipe for posting frequency, but one rule that should always apply is: if you don't have anything of value to share, don't say anything at all. Practices should start by sharing 3 times a week. Check your interaction, and make adjustments as needed. If you increase posting frequency, and the level of interaction with your followers does not decrease, you are in good shape. With that said, let's be sure to keep this in perspective. By having someone follow or like your social media site, you have been given the honor of having access to the stream of information that sites share with your follower on a daily (hourly) basis. In their stream or newsfeed, you are listed with many other brands, organizations, and causes. Finding the balance between being topical and a nuisance can be difficult. Ease into the process, and keep an eye on how many people unlike your sites on a monthly basis.

Relevance

As mentioned above, please remember that the person receiving your message voluntarily followed your practice with the understanding that you would be providing information about your practice and caring for pets. Deviating from the expected subject matter can be offensive and will cause followers to unlike your sites.

Determining what to write on any specific day requires us to revisit the posting calendar. Your practice's social media posting calendar should be developed by examining the following:

- Practice Marketing Plan (considering monthly promotions or educational focus)
- Pet focused days/months that are promoted on a national basis (mentions in news media will support your posts)
- National non-pet holidays/events/milestones (create a pet spin on the event) Local events or concerns that affect pet owners
- Local personalities who can contribute to topics

With the above guides in place, you can develop annual, monthly, and weekly themes. These themes should be listed in a calendar form allowing for posts to be written and developed over time. Time will allow your practice to gather pictures, to recall real stories to support the message, to identify clients who can offer testimonial support, and even to alert the media that you will be discussing those matters (we'll discuss the media in a later lecture...but this is all connected).

Where Do I Get Content?

Great question. The best scenario is for you to create it yourself, but while this sounds great, the reality of day to day practice life might cause social media content creation to slide to the back burner. Some practices are large enough to hire dedicated marketing staff members who can handle this task, but those situations are unreasonable for most.

Many practices seek professional assistance in creating content. Make certain that the firm creating the content is dedicated to the veterinary space and is creating CUSTOM content for you. Use of generic or scraped content can be very dangerous, often leading to negative results. The pitfalls of using this content are that Google considers this practice to be duplicate content.

Because Google is always looking to credit the original author of the content, it gets confused and actually penalizes websites and social media channels for using content published on other sites, because it looks at the practice as scraping content or stealing content that is not owned by the site. If you would like to learn more, simply Google the term duplicate content, and you'll see what I mean. Besides being penalized (which is a huge deal), why would you want to look like or sound like every other veterinary practice? Your message should be unique. A professional content provider should consult with you on a monthly basis, learn about your practice's needs, and provide custom content that aligns with your SEO plans.

Combining this activity with your practice interaction on the sites provides a very stable environment. Before you think that you cannot afford such a service, this service can be found for around $200-300 a month. What would you pay a staff member to perform such a task?

Keyword Stuffing

So, you are armed with your keyword and key phrase list. The goal is to get as many of those words/phrases into every paragraph as possible, right? Nope. You need to develop a balance considering frequency of usage, positioning of the text, and the relationship these factors have on the search engine algorithm. Professional assistance may be helpful here. Reputable firms will have the ability to analyze content and determine the optimal blend. The rule of thumb is to always write for the person you are trying to affect. Do so with the search engines in mind, but let the user experience lead the way. How much respect would you have for a practice that wrote a blog that was stuffed with keywords and phrases?

Key Takeaways

- Understand the relationship between content used in social media and in search engines.
- Know which keywords and key phrases are being used to affect practices.
- Develop a posting calendar and come up with content well in advance.
- Write for the user experience; Google will follow good content.
- Avoid duplicate content. Be as custom as possible.

Building and Growing Your Exotic Animal Practice

Dan H. Johnson, DVM, Dipl ABVP (Exotic Companion Mammal)

Session #345

Affiliation: From Avian and Exotic Animal Care, 8711 Fidelity Blvd, Raleigh, NC 27617, USA.

Exotic animals represent about 20% of the companion animal market for veterinary services.[1] Pet owners readily seek out veterinary care for birds, small exotic mammals, reptiles and amphibians. The greatest obstacles to growing your exotic animal practice are: (1) helping potential clients find you and (2) convincing them that your training and expertise matter.

Marketing

Properly naming your practice plays an important role in building public confidence. To do so, a stand-alone exotic animal hospital will usually incorporate the terms avian, bird, or exotic into the practice name. Similarly, a companion animal hospital that offers competent avian and exotic animal care should distinguish that service with a name that effectively communicates the level of expertise offered (e.g., Westover Animal Hospital and Exotic Animal Clinic or Pet Bird and Exotic Animal Service of Westover Animal Hospital). Having a separation of the exotic practice from your dog and cat practice is less threatening to your referral base, is more appealing to your exotic-animal clientele, and allows you to position yourself as the avian and exotic animal expert in your market.

Let your current clients know that you see exotic animals. Your practice logo should communicate that birds, reptiles, and small exotic mammals are seen at your hospital. The reception area should contain artwork and products that reflect your interest in exotic animal medicine. Place exotic animal brochures and client education materials in the exam room and on the countertops. There should me reminders that you treat exotic animals throughout the office. Membership plaques from the Association of Avian Veterinarians (AAV), Association of Reptilian and Amphibian Veterinarians (ARAV), and Association of Exotic Mammal Veterinarians (AEMV) should be mandatory.

Your exam room computer can passively communicate to clients which species you treat and what services you offer. Place a number of your best, most interesting photographs together in a single folder, and use the computer's screensaver to play a slideshow of those photographs. This also works on a large, flat-screen television placed in your waiting room, and most current models will play your slideshow from a flash drive (no external computer required). You need nice cards and brochures, but a refrigerator magnet never, ever gets discarded. An attractive refrigerator magnet will live on for many years!

Setting Appointments

Triage for exotic animals requires skill. In general, birds and exotic animals are much sicker than dogs and cats are before they begin to show signs of disease. Compounding this, many exotic pets do not get the daily close attention that traditional pets do. Often, exotic animal owners have had to call 2-3 veterinary practices before they find one that will treat their pet. As a result, a high percentage of avian and exotic cases present as emergencies.

Staff should be trained to properly triage exotic animal cases and schedule these pets accordingly. Most sick animal calls should be seen the same day as work-in emergencies. A large percentage of surgical and medical cases will not materialize if treatment appointments are delayed until the next workday. Exotic animals are inconvenient to treat, and if time will not allow an emergency same-day appointment, a referral should be offered.

Exotic Animal Hospital Setup

Exam rooms should be small, with a low ceiling, and lights that can be dimmed. It is best if there are no windows, or high places such as planters for birds to escape to. Exam rooms need to be ferret-proofed to prevent small animals from getting beneath and behind your base cabinets. Check the space under the door to make sure a small rodent or ferret couldn't possibly sneak through. Given the opportunity, they will. Birds, rabbits, and small exotic mammals are very sensitive to noise. In the hospital wards, limit noise due to barking and other things to reduce patient stress. Create a visual barrier between prey animals and predatory species including dogs and cats. Reduce drafts from air conditioning vents. Choose the warmest part of the building for birds and exotic animals. The hospital will also need a sink or tub large enough for waterfowl and iguanas to swim in. The isolation ward should be dark, warm, and away from all noise. Under ideal circumstances, there should be an incubator with a nebulizer, oxygen, heat, and humidity control.

Sharing What You Know

Many avian and exotic practitioners are reluctant to share knowledge with local colleagues. This is understandable since it took time, money and effort to gain all of the exotic animal knowledge and expertise that you have acquired. After all, "Why should I be the CE for some vet who calls when he has a rabbit spay on the table and realizes he has a question?" But this author believes that taking calls from area practitioners is one of the best ways to show neighboring veterinarians how much more you know than they do. Look at every phone call from a veterinarian as a chance to earn a future referral. Answering a drug dose question for a colleague is likely to earn you a medicine or surgery referral in the future. If you are not willing to come to the phone, how else is the neighboring vet community going to learn how much you know? If you think that sharing a little info with a colleague will take away from your business, then you vastly underestimate the barrier to competition than we enjoy as exotic animal practitioners. Sharing a dose or technique with a colleague will help you both.

Showcase Your Exotic Animal Practice

If you are trying to grow your exotic animal caseload, then invite everyone back to see what you do. Ask your clients behind the scenes to check out the facility and see you working. Say "yes" to school tours.

Pre-veterinary and veterinary students are your future referral base. Host as many as you can! Roll out the red carpet for the veterinarians and staff at other veterinary hospitals. Allow interested veterinarians to come over and get a taste for what it is you do.

Getting the Word Out

Letting the public know that you see birds and exotic animals is half the battle. Of course you need a good website and an active Facebook page. But, should you advertise? The author has chosen to only advertise in highly targeted areas such as the state veterinary medical association quarterly newsletter because it goes directly to other area veterinarians. No magazines or newspapers, and no coupon packs because the general public is not targeted enough. A few ways that the author markets his practice directly to his target audience are:

1. Being active in, an exhibitor at, and a sponsor of area exotic animal events. Examples from the author's local geographic area: reptile and exotic animal show, aquarium society meetings, koi club meeting, backyard chicken enthusiasts annual home show, bird club meetings, ferret club and rescue events, rabbit meetup and rescue events, museum events (reptile and amphibian day, bug festival), annual turtle festival at local park;

2. Speaking at schools, libraries and summer camps (especially your client's kids!);

3. Speaking and writing in veterinary meetings/publications (getting your name on anything that goes to the general vet referral base is excellent!);

4. Staying active with the veterinary school; and

5. Teach exotic animal medicine at the area technician school, hosting technician interns, and helping technician candidates who attend school online.

References

1. American Pet Products Association. 2015-2016 APPA National Pet Owners Survey Statistics: Pet Ownership & Annual Expenses. Available at: http://www.americanpetproducts.org. Accessed July 6, 2016.

Setting Up a Private Practice Residency

Angela M. Lennox, DVM, Dipl ABVP (Avian), Dipl ABVP (Exotic Companion Mammal), Dipl ECZM (Small Mammal)

Session #346

Affiliation: From Avian and Exotic Animal Clinic, 9330 Waldemar Road, Indianapolis, IN 46268, USA.

Definition of Residency in Medicine

In the medical field, a residency is defined as a "period of advanced medical training and education that normally follows graduation from medical school and licensing to practice medicine and that consists of supervised practice of a specialty in a hospital and in its outpatient department and instruction from specialists on the hospital staff." The concept of the residency program was inaugurated in the medical field 127 years ago. The American Veterinary Medical Association (AVMA) received applications for the very first two veterinary specialties in 1950; these were the American College of Veterinary Pathologists and the American Board of Veterinary Public Health. AVMA created the Advisory Board on Veterinary Specialties in 1959, which became the American Board of Veterinary Specialists (ABVS) in 1992. ABVS is the umbrella organization for US veterinary specialties, and is composed of one member from each AVMA-recognized specialty plus other advisors. ABVS regulates current veterinary specialties, and evaluates applications for newly created specialties.[1]

Pathways for Private Practice Residency in the United States

The American Board of Veterinary Practitioners (ABVP) was established in 1978 as a means for veterinarians in private practice to obtain specialist status without the burden of completing a concentrated residency. Requirements match the natural progression of private practice, emphasizing years of clinical practice along with publication and testing requirements.

ABVP offers a residency pathway as well as private practitioner pathway. Specialties of interest to exotic practitioners include Avian, Exotic Companion Mammal and Reptile and Amphibian. There are currently 18 ABVP-approved residency sites in Avian Medicine (7/18 at Universities) and 7 in Exotic Companion Mammal (3/7 at Universities). Not every site may offer a residency each year. A Residency Program Guide provides instructions and guidelines for establishment of a residency program.[2]

The American College of Zoological Medicine established a specialty program in 1983, which was approved by ABVS in 1988. A residency program was initiated in 1992. A residency emphasizing Zoological Companion Animals is available. To the best of the author's knowledge there is no private practice ACZM Companion Animal Residency.

Intern and Residency Matching Programs

The Veterinary Intern and Residency Matching Program (VIRMP) is sponsored by the American Association of Veterinary Clinicians and was established as a matching program for candidates seeking intern and residency opportunities and both University and private practice programs. VIRMP follows similar principles to those adopted by the Physicians National Intern and Resident Matching Program. According to VIRMP:

"The applicant completes and application packet and submits the application via this website. The VIRMP uses a central computer facility for matching prospective interns and residents with programs. Applicants create a rank order list of programs to which they want to apply. Institutions and private practices create a rank order list of applicants who they are willing to accept for their program. Both rank order lists are confidential. The VIRMP utilized sophisticated computer software to match applicants and programs in a way that insures each receives the highest available option on their rank order lists."

Internship programs are not regulated by any entity; however, AVMA has developed the Veterinary Internship Guidelines to "establish reasonable expectations for internship providers as well as participants." Despite the guidelines, there is no governing board; therefore, the structure, goals and mentorship are entirely self-regulated by the program offering the internship. For this reason, quality varies significantly leading to the warning: buyer beware. The assumption is that the intern will receive less compensation in exchange for more intense, focused mentorship; however, some internships have earned a reputation for deficient or absent mentorship and appear to be a way to obtain cheap labor.

Residencies are regulated by the sponsoring organizations. The VIRMP lists residencies in the categories of Ambulatory, Anesthesia, Avian Medicine, Behavior, Cardiology, Clinic Nutrition, Dentistry, Dermatology, Exotic/Wildlife/Zoo, Poultry Medicine, Radiation Oncology/Therapy, Small and Large Animal Surgery, Theriogenology, and more.[3]

Obligations of Private Practice Residencies

Residency training programs in private practice present distinct advantages and disadvantages over university-based programs. Advantages include the benefits of a higher caseload, in some cases significantly higher. Disadvantages are the lack of exposure to the resources of a university, including access to multiple specialists in a single location.

Prior to application for residency, the prospective mentor must consider the following:

1. Will the mentor be able to provide significant contact hours with the resident?
2. Will the mentor be available by cell phone or electronic communication nearly 24 hours per day, 7 days per week when not directly mentoring?
3. How will the mentor provide planned, direct teaching moments?
4. How will the resident gain exposure to specialists outside the mentor's specialty?
5. How will the mentor objectively evaluate the resident during the course of the residency?

Modifications to the routine of private practice may be significant when mentoring a resident. These may include adjusting schedule to ensure residents are alone in clinics as seldom as possible, the establishment of planned staff training, case rounds and journal clubs. All of these may represent a significant paradigm shift and change in priorities for the mentor.

References

1. American Veterinary Medical Association. American Board of Veterinary Specialties. Available at: www.avma.org/ProfessionalDevelopment/Education/Specialties/Pages/default.aspx. Accessed July 6, 2016.

2. American Board of Veterinary Practitioners. Residency Program Guide. Available at: http://www.abvp.com/why-residency. Accessed July 6, 2016.

3. American Association of Veterinary Clinics. Veterinary Internship and Residency Matching Program. Available at: www.virmp.org. Accessed July 6, 2016.

Updates on Veterinary Drug Compounding

Lance M. Roasa, DVM, MS, JD

Session #347

Affiliation: From Superior Veterinary Care, Lincoln, NE 68504, USA.

Abstract: Veterinary practitioners use compounded medications extensively, but the legal status of the drug is often poorly understood. The Food and Drug Administration's (FDA) regulations don't allow compounded from bulk ingredients, which is necessary for many medications treating commonly presented conditions. The FDA has released draft guidance, which help outline new thinking on the status of compounding legally. A best-practices approach will help exotics practitioners stay up to date on FDA publications, know the legal status of the drugs they use in practice, and reduce liability and adverse events.

What is a Drug?

Although veterinarians use the term drug in common language, technically it is a legally defined term with a very broad definition. The legal definition of drug is found in the Food Drug and Cosmetic Act of 1938 (FD&C Act) as "articles intended for use in the diagnosis, cure, mitigation, treatment, or prevention of disease in man or other animals."[1] Further broadening the definition Congress also names drugs as "articles (other than food) intended to affect the structure or any function of the body of man or other animals."[2] This expansive definition encompasses virtually every substance that a veterinarian gives or prescribes, other than food or water and brings them under the jurisdiction of the Food, Drug and Cosmetic Act with regulation and enforcement of the Food and Drug Administration (FDA).

After defining the term drug, the FD&C Act mandates that all animal drugs that are new to the market undergo the New Animal Drug Application (NADA).[3] The NADA process is a long and potentially expensive process that tests the new animal drug for safety, effectiveness, the manufacturing process, and stability among other characteristics. Once a drug is tested and is approved through the process and also is protected by a patent, it is called a legend (or pioneer) drug. A pioneer drug must be labeled with the statement (or legend): "Caution: Federal law restricts this drug to use by or on the order of a licensed veterinarian."[4] When a veterinarian reaches for a drug that bears this label, and they follow the label requirements, they can rely on the fact that the contents have been tested, approved and manufactured in accordance with the FDA's regulations.

What is a Generic Drug?

The FD&C Act contains further specific requirements for generic drugs. After a drug has cleared the NADA process and is not protected by a patent, another manufacture can submit for the Abbreviated New Animal Drug Application (ANADA) process.[5] The ANADA bypasses some of the safety and efficacy testing that a New Drug must undergo and concentrates on the manufacturing processes of the generic manufacturers.[6] To carry the term generic drug, the drug must undergo and pass the ANADA process. Thus, a compounded drug that is not ANADA approved is not a generic drug.

The term generic name also is different from generic drug. A drug's generic name is its established name, which is also usually the active ingredient.[7] The word generic can easily be confused with a drug that has been sponsored through the ANADA process. There are surprisingly few true generic drugs. The public can access drugs that are approved as pioneer drugs or generic drugs the FDA's database.[8]

What is Extralabel Use?

According to the FD&C Act every drug that does not bear the legend default is deemed unsafe and is adulterated or illegal to administer and or prescribe.[9] Every time a drug is used not according to label requirements on the legend drug, it is said to be extralabel use. Unfortunately, veterinary practice is not that black and white. In most veterinary practices the patients don't always meet the label's requirements for size, species and condition. Very few drugs for exotic species have undergone the NADA process, because the return on investment for the drug sponsor (manufacturer) is minimal. In an attempt to solve these grey areas of practice, the Animal Medicinal Use Clarification Act (AMDUCA) was enacted in 1994. AMDUCA allows veterinarians, with a valid veterinary client patient relationship (VCPR) to use drugs in an extra label manner if the FDA regulations are followed.[10] Those regulations define extralabel use as:

> "Actual use or intended use of a drug in an animal in a manner that is not in accordance with the approved labeling. This includes, but is not limited to, use in species not listed in the labeling, use for indications (disease and other conditions) not listed in the labeling, use at dosage levels, frequencies, or routes of administration other than those stated in the labeling, and deviation from labeled withdrawal time based on these different uses."
>
> Title 21 Code of Federal Regulations, part 530.3(a)

This regulation gives veterinarians wide leniency to use drugs in extralabel manner in their practice with as long as human safety in the form of withdrawal times is not compromised. This does not allow the drug to be adulterated or mixed in a manner that is not on the label.

Where Does Compounding Come In?

Even with wide ability to use drugs in an extralabel manner, there are still species and conditions that cannot be treated without additional measures. This may mean mixing the drug in a more dilute or stronger concentration, adding flavoring or an additional substance, or changing the formulation of the medication (i.e., grinding a pill into a power then liquefying). The FDA regulation allows this, under strict guidance. The animal compounding regulations give these requirements, among others, for compounding under AMDUCA:

- There is no approved animal or human drug that can be used for the condition diagnosed under labeled or extralabel uses,
- a human drug cannot be used when an animal drug is available,
- the compounding is done by a licensed veterinarian or pharmacist,
- adequate procedures are followed to ensure safety and effectiveness, and
- it is in compliance with applicable state laws.

Most importantly, the regulations clearly state, "Nothing in this part shall be construed as permitting compounding from bulk drugs."[11] A bulk drug is a means the starting point is an active pharmaceutical product (API) and not product that has gone through the NADA, ANADA process or the human equivalent. In other words, a bulk drug is the pure compound, without the inert ingredients for stability, absorption, taste or other reasons in pills, liquids and other consumer ready drugs.

Currently, according to the FD&C Act and FDA regulation, compounding can be performed in very controlled circumstances, and not starting with a bulk drug. Still practitioners are faced with scenarios that require a bulk starting point. Many times a legend or generic drug contains too many inert ingredients or is not at a concentration that is conducive to compounding for very small or very large patients. Some drugs that veterinarians reach for have no legend or generic preparations at all. The prohibition on bulk starting points seems to fly in the face of what is happening in everyday practice and pharmacies all over the country. Compounding pharmacies seem to offer every medication imaginable, in every concentration and administration route, and many are made from bulk-sourced drugs. Further confusing the topic of compounding is differing laws and regulations in every state and selective enforcement of the rules by the FDA and state agencies that oversee pharmacies and veterinary practices. State laws vary on the stocking, dispensing and resale of compounded medications.

What about State Laws and Regulations?

State laws vary greatly on the issue of compounded formulations, and a practicing veterinarian should check with local sources before relying on this data. The state pharmacy boards are typically concerned with the licensing and regulation of pharmacists, and most states have specific requirements regarding veterinary drug compounding. The American Veterinary Medical Association (AVMA) has compiled these resources. An issue with an extremely wide range of variance at the state law level is whether a veterinarian can dispense or resale compounded drugs. Two states, New York and New Mexico, do not allow compounded drugs to be used, either in office or as dispensed.[12] As of January 2016, 21 states had specific provisions allowing in office use, but prohibiting dispensing or reselling.[13] Nine states specifically allow administration in office-use and for dispensing to clients.[14] Twelve states specifically allow administration of a compounded medication but are not clear on dispensing or reselling.[15] The remaining states have unclear laws.[16] On a national level, this has contributed to confusion as to when and how a veterinarian can administer and dispense compounded drugs. A practitioner must look to local resources for the laws and regulations for compounded medications.

What about MUMS?

The Minor Use/Minor Species Act of 2004 (MUMS) was intended to make more medications available to veterinarians treating minor species and uncommon diseases of the major species.[17] Minor species are defined all animals other than humans, horses, dogs, cats, cattle, pigs, turkeys, and chickens. MUMS modifies the FD&C Act in three categories; conditional approval, designation and indexing.

Conditional approval

With a Conditional Approval under the MUMS act a sponsor can make the drug available before full efficacy testing, but after proving safety. Conditional approval speeds the drug to market, as long as there is a reasonable expectation of effectiveness. The drug can stay on conditional approval up to 5 years, while effectiveness data is being collected.

Designation status

Designation provides drug sponsors incentives for approval with grants for safety and effectiveness testing and provides 7 years of exclusive marketing once approval is received. Currently 137 drugs are on the designation list.

Indexing

An Index drug is for drugs that are use in species that are too rare or too varied to be subjected to well-controlled studies. The FDA maintains a list, the Index of Legally Marketed Unapproved New Animal Drugs for Minor Species (the Index), of drugs that can be used without full approval. Currently the Index lists 11 drugs for unapproved usage.

Legally, Where is Compounding Heading?

In May of 2015, the FDA release draft guidance for industry (GFI #230) addressing compounding from bulk drugs.[18] Although the document is not finalized and it is not regulation, it does outline the FDA's current thinking and when they will take action regarding compounded drugs. GFI #230 contains 11 steps to follow when prescribing drugs from compounding pharmacies.

1. The compounding is to be done at a state licensed pharmacy.

2. A valid prescription identifies an individual animal, and the compounding is done after the receipt (unless specific guidelines are met).

3. The drug is not intended for use in food-producing animals, and the prescription or accompanying paperwork contains the statement: "This patient is not a food-producing animal." Chickens, swine, sheep and goats are always considered food-producing regardless of whether the individual is a family pet. This means that this guidance cannot apply to pot-bellied pigs or pet-chickens or goats.

4. To compound from a bulk substance, the veterinarian has to expect a clinical difference from a compounded substance from an FDA-approved starting point. Also, the prescription or accompanying paperwork has to specify why that clinical difference exists.

5. If an FDA-approved starting point exists, then the pharmacist has to make a determination the desired compounded drug cannot be made from the FDA-approved drug.

6. The prescription or accompanying paperwork must state the animal's species and the statement:

   "There are no FDA-approved animal or human drugs that can be used as labeled or in an extra-label manner under section 512(a)(4) or (5) and 21 CFR part 530 to appropriately treat the disease, symptom, or condition for which this drug is being prescribed."

7. The bulk drug source has to be manufactured by a registered establishment, and a valid certificate of analysis has to accompany the bulk drug.

8. The drug has to be compounded following USP guidance, and sterile drugs have to be prepared in USP ISO Class 5 standards.

9. The drug is not sold or transferred by any entity other than the compounding pharmacy. A veterinarian can administer the drug, but not resale or transfer the drug.

10. Within 15 days of becoming aware of an adverse event, the pharmacy has to report to the FDA.

11. The label contains the species and name of the patient and the name of the owner or caretaker of the animal patient.

What are Best Practices for Using Compounded Drugs?

It is the author's opinion that following these FDA-outlined steps when reaching for a compounded medication will not draw the scrutiny of the FDA. Further, additional steps should be taken to ensure a best-practices approach to compounded medications to ensure maximal patient safety and liability protection.

1. Know the legal status of the drug you are administering or prescribing.
   - Is the drug an FDA-approved animal drug? If so, are you following label instructions?
   - Is the drug a generic drug that has undergone the ANADA process?
   - Is the drug a compounded drug, but under an AMDUCA exemption?
   - Is the drug compound from a bulk drug source but you deem it is necessary?

2. Only use extralabel and compounded drugs when medically necessary.

3. Document the legal status of the drug in the medical record.

4. Consult with the client about the legal status of the drug, its safety, efficacy, and potential for adverse events.
   - Obtain the client's consent to use of extralabel and compounded drugs, as well as their understanding of safety and adverse events.
   - Document the client's consent and understanding in the medical record.

5. Ensure follow-up regarding the patient's response to extralabel and compounded drugs.
   - Document the response, any adverse events, and/or owner compliance.
   - Report adverse events to the pharmacy.

6. Know your compounding pharmacist(s).
   - Ensure the compounding pharmacy you are using is Pharmacy Compounding Accreditation Board (PCAB) accredited.[19]
   - Talk to your compounding pharmacists, ask questions (e.g., how the prescription is made, bulk or FDA-approved product?), and check the certificates of analysis.
   - Understand their protocols for sterility and shipping in hot and cold weather.
   - Confirm that they carry liability insurance that will cover adverse events.

7. Know and follow the beyond-use-date (BUD); know how the BUD was derived.

8. The use of compounded formulations in animals to be used for food is not allowed.

9. Know your state laws regarding stocking, reselling, dispensing, administering, and/or labeling compounded medications.

10. Stay informed of policy changes, and watch for FDA regulation changes and safety/efficacy data.

References

1. Title 21 US CODE § 321(g)(1)(B).

2. Title 21 US CODE § 321(g)(1)(C).

3. Title 21 US CODE § 360b.

4. Title 21 Code of Federal Regulations, part 201.105(b)(1).

5. Title 21 US CODE §§ 360b(b), (c).

6. Title 21 US CODE § 360b(c)(2)(A).

7. See Generic Animal Drugs: Approved or Unapproved? Available at: http://www.fda.gov/AnimalVeterinary/GuidanceComplianceEnforcement/ComplianceEnforcement/UnapprovedAnimalDrugs/ucm249392.htm. Accessed May 9, 2016.

8. Animal Drugs @ FDA Last Accessed May 9 at 2016 Available at: http://www.accessdata.fda.gov/scripts/animaldrugsatfda/index.cfm?gb=1. Accessed July 6, 2016.

9. Title 21 US CODE § 360b(a)(1).

10. Title 21 US CODE § 360b(a)(4).

11. Title 21 Code of Federal Regulations, part 530.13.

12. American Veterinary Med Assoc. State Summary Report. Aug 2015. Available at: https://www.avma.org/Advocacy/StateAndLocal/Pages/compoundinglaws.aspx. Accessed January 28, 2016.

13. State laws and regulations that allow veterinary offices to administer compounded products but specifically prohibit them from dispensing or reselling products compounded by a pharmacy: Alabama, Alaska, Arizona, Arkansas, Connecticut, Georgia, Iowa, Kentucky, Louisiana, Maryland, Missouri, North Carolina, North Dakota, Ohio, Oklahoma, Rhode Island, Utah, Vermont, Washington, West Virginia, Wyoming. Available at: https://www.avma.org/Advocacy/StateAndLocal/Pages/compoundinglaws.aspx. Accessed January 22, 2016.

14. State laws and regulations that allow veterinary offices to administer compounded products and dispense the products to clients but may be subject to conditions or limitations in some cases: California, Florida, Maine, Michigan, Minnesota, Nebraska, Tennessee, Texas, Virginia. Available at: https://www.avma.org/Advocacy/StateAndLocal/Pages/compoundinglaws.aspx. Accessed January 22, 2016.

15. State laws and regulations that allow veterinary offices to administer compounded products but do not specifically address or the law is not clear regarding dispensing products compounded by a pharmacy: Colorado, Delaware, Idaho, Illinois, Massachusetts, Mississippi, Montana, Nevada, New Hampshire, New Jersey, Oregon, South Carolina. Available at: https://www.avma.org/Advocacy/StateAndLocal/Pages/compoundinglaws.aspx. Accessed January 22, 2016.

16. States where with no laws and regulations addressing compounding: Hawaii, Indiana, Kansas, Pennsylvania, South Dakota, Wisconsin. Available at: https://www.avma.org/Advocacy/StateAndLocal/Pages/compoundinglaws.aspx. Accessed January 22, 2016.

17. Title 21 US CODE §§ 360ccc.

18. Draft Guidance for Industry #230, Compounding Animal Drugs from Bulk Substances, May 2016, Available at: http://www.accessdata.fda.gov/scripts/animaldrugsatfda/index.cfm?gb=1. Accessed July 6, 2016.

19. Pharmacy Compounding Accreditation Board, Find a PCAB Accredited Pharmacy Available at: http://www.pcab.org/accredited-locations.html. Accessed May 9, 2016.

Gap in pagination due to formatting issues.

Pages 37-684

Section 24

ARAV Plenary Sessions

Susan Horton, DVM
ARAV President-Elect
Moderator

How Veterinarians Help in the Recovery of Two Extinct Galápagos Tortoises (*Chelonoidis* sp.)

Paul M. Gibbons, DVM, MS, Dipl ABVP (Reptile and Amphibian)

Session #329

Affiliation: From the Turtle Conservancy, 1794 McNell Rd, Ojai, CA 93023, USA.

Lonesome George, the last remaining Pinta Island Tortoise died in 2012, and later that year scientists published their plan on how to bring his species back from extinction and save the ecosystem on his native island in the Galápagos archipelago. In November 2015, 3 veterinarians joined a team of over 60 scientists, field biologists, park rangers, pilots, and sailors on what has been called the most ambitious extinct species recovery effort ever attempted in the world. The team used machetes to carve their way through 20 square miles of rugged wilderness covering the western slope of Vulcán Wolf on Isabela Island in the Galápagos, where an unusual assemblage of tortoises, including descendants of Lonesome George, is thriving. These tortoises likely wound up here sometime in the 19th century after they had been collected on neighboring islands to provision whaling ships, then dumped overboard to reduce weight. Those that managed to swim ashore settled on the island and bred with the native tortoise species. The author encountered more than 500 wild tortoises while he was there. The expedition chose 32 tortoises with the best characteristics for captive breeding, which will help restore extinct species and the tortoise-dependent ecosystems on the islands of Floreana and Pinta.

Magnetic Resonance Imaging in *Trachemys* and *Pseudemys*: Dimensions, Signal Intensities and Positions of Select Organs

K. Mathes, DVM, Dr. med. vet., Cert Spec in Reptiles and Amphibians, Dipl ECZM (Herpetology),
M. Schnack, DVM, Dr. med. vet.,
M. Fehr, DVM, Dr. med. vet., PhD

Session #156

Affiliation: From the Clinic for Small Mammals, Reptiles and Birds, University of Veterinary Medicine Hannover, Buenteweg 9, D-30559 Hannover, Germany.

Although the shell of chelonians restricts radiologic and ultrasonographic visualization of inner organs, magnetic resonance imaging (MRI) is an excellent tool to illustrate dimensions and positions of organs within the coelomic cavity of these animals. Organs of clinically healthy red-eared sliders (*Trachemys scripta elegans*), yellow-bellied sliders (*Trachemys scripta scripta*), coastal plain cooters (*Pseudemys concinna concinna*), and hieroglyphic river cooters (*Pseudemys floridiana*) of different sizes and weights and both genders were examined in dorsal, sagittal and transverse planes using MRI. For MR investigations, all turtles were sedated and positioned in the same manner with maximally extended neck and extremities. Reference data were established for dimensions and locations of inner organs such as heart, liver, gallbladder and kidneys based on T1-weighted images. Ratios were used to take the specific impact of the varying volumes and weights of the turtles into account and thereby generalize the reference data to act as a guide for subsequent studies. Measurements of signal intensities of kidney and liver were performed as well and compared to signal intensities of muscles to serve as reference in further examinations in North American freshwater turtles. The applied methods of measurement and evaluation of the organs can also be used as a template for future studies in other turtle species.

Section 25

ARAV Infectious Diseases

Marike Visser, DVM;
Daniella Inman
Moderators

Experimental Inoculation of Ball Pythons *(Python regius)* with Snake Arenaviruses

David Sanchez-Migallon Guzman, LV, MS, Dipl ECZM (Avian),
Dipl ECZM (Small Mammal), Dipl ACZM,
Mark D. Stenglein, PhD,
Michelle Hawkins, VMD, Dipl ABVP (Avian),
Valentina Garcia,
M. Kevin Keel, DVM, PhD, Dipl ACVP,
Tracy Drazenoich, DVM,
Joseph L. DeRisi PhD

Session #291

Affiliation: From the Department of Veterinary Medicine and Epidemiology, School of Veterinary Medicine, University of California, Davis, CA 95616, USA (Guzman, Hawkins, Drazenovich), Department of Microbiology Immunology and Pathology, College of Veterinary Medicine and Biomedical Sciences, Colorado State University (Stenglein), Department of Pathology, Microbiology and Immunology, School of Veterinary Medicine, University of California, Davis, CA 95616, USA (Keel), Department of Biochemistry and Biophysics, School of Veterinary Medicine, University of California San Francisco, San Francisco, CA 94158, USA (Garcia, DeRisi).

Inclusion body disease (IBD) is a fatal disease of snakes. IBD is most commonly diagnosed in boas and pythons, but has been reported in several other species. Disease manifestation is variable in different species.[1] Snake arenaviruses have been associated with IBD in boids,[2,3] but causality has not yet been demonstrated. For this study, 4 ball pythons (*Python regius*) that tested negative on polymerase chain reaction (PCR) technique for snake arenavirus from blood and liver biopsies were used. Two ball pythons were inoculated with 2 different viral strains of snake arenavirus while 2 additional ball pythons served as a control. The viral strains used to inoculate the 2 snakes were isolated from previous boa constrictors (*Boa constrictor*) with inclusion body disease. The infected snakes developed severe neurologic clinical signs approximately 9 weeks after inoculation and had moderate encephalomyelitis, meningitis and ganglioneuritis on histopathology. The presence of viral RNA for snake arenavirus was confirmed and quantified by qRT-PCR (real-time quantitative reverse transcription PCR) in the infected snakes and immunofluorescence was performed in multiple tissues demonstrating the presence of the virus in the brain. The control snakes remained without clinical signs during the same period and no significant lesions were found on histopathology. There was no snake arenavirus RNA found in the control snakes and immunofluorescence was negative. The results of this study establish a causative relationship between snake arenavirus and IBD in ball pythons.

Acknowledgments: This study was supported by the Reptile Research Fund at the University of California Davis and the Howard Hughes Medical Institute.

References

1. Chang L-W, Jacobson ER. Inclusion body disease, a worldwide infectious disease of boid snakes: a review. *J Exot Pet Med.* 2010;3:216-225.

2. Stenglein MD, Sanders C, Kistler AL, et al. Identification, characterization, and in vitro culture of highly divergent arenaviruses from boa constrictors and annulated tree boas: candidate etiological agents for snake inclusion body disease. *MBio* 2012;3:e00180-00112.

3. Stenglein MD, Jacobson ER, Chang LW, et al. Widespread recombination, reassortment, and transmission of unbalanced compound viral genotypes in natural arenavirus infections. *PLoS Pathog* 2015;11:e1004900.

Immune response and Presence of Viral RNA in Tissues of Hatchling American Alligators *(Alligator mississippiensis)* after Vaccination with a Killed West Nile Virus Vaccine

Christine T. Higbie, DVM,
Javier G. Nevarez, DVM, PhD, Dipl ACZM, Dipl ECZM (Herpetology),
Alma F. Roy, MS, PhD,
Fabio Del Piero, DVM, PhD, Dipl ACVP

Session #124

Affiliation: From the Department of Veterinary Clinical Sciences (Higbie, Nevarez) and the Department of Pathobiological Sciences (Roy, Del Piero), School of Veterinary Medicine, Louisiana State University, Baton Rouge, LA 70803, USA.

West Nile virus (WNV) can cause mortalities in captive reared American alligators. A killed WNV vaccine is commercially available (Vetera WNV, Boehringer Ingelheim Vetmedica, Inc., St. Joseph, Missouri, 64506, USA). A case from an alligator farm in Louisiana revealed that tissues of vaccinated animals were positive for WNV by reverse transcriptase polymerase chain reaction (RT-PCR) while non-vaccinated animals were negative, suggesting the RT-PCR might detect the vaccine in the tissues. The vaccine product itself was positive for WNV via RT-PCR. This study was designed to determine if vaccination could lead to positive RT-PCR from tissues and also to evaluate the antibody response. Eighty-one, 7-day-old hatchlings were vaccinated with 0.25 ml WNV vaccine into the lateral tail musculature and 26 control alligators were injected with 0.25 ml sterile water into the lateral tail musculature on days 1 and 14. Blood was obtained daily from 3 animals in the treatment group prior to euthanasia and necropsy until all animals were euthanatized (27 days). Thirteen of the control alligators were sampled on day 14 and the remaining were sampled on day 27. WNV titers were obtained via serum neutralization. Real time RT-PCR was performed on the brain, liver, kidney, and tail musculature from each animal. Preliminary results revealed antibodies in both the control and treatment groups but viral RNA was only identified in the treatment group. The results confirm that RT-PCR can detect viral RNA from the vaccine in tissues of vaccinated alligators. The presence of antibodies in the control group without any mortalities suggests maternal transfer of antibodies.

Polymerase Chain Reaction Detection of Nidoviruses in Live Pythons and Boas

Rachel E. Marschang, PD Dr. med. vet., Dipl ECZM (Herpetology), FTÄ Mikrobiologie, ZB Reptilien, Ekaterina Kolesnik, TÄ

Session #135

Affiliation: From Laboklin GmbH & Co. KG, Steubenstr. 4, 97688 Bad Kissingen, Germany.

Viruses of the order Nidovirales have recently been detected in 2 species of pythons (ball pythons, *Python regius*, and Indian rock pythons, *P. molurus*) in the United States and Europe. They are believed to be an important cause of respiratory disease in these animals. A conventional polymerase chain reaction (PCR) was used to screen clinical samples from 201 live pythons and boas for nidoviruses (95 pythons, 84 boas, and 22 snakes of unknown species). Nidoviruses were detected in 30 of the 201 snakes examined (14.9%). This included 26 pythons (27.4% of the pythons tested), 2 boas (2.4% of the boas tested), and 2 snakes of unknown family. There was a highly significant difference in the percentage of positive pythons and the percentage of positive boas. Viruses were detected in ball pythons (16 positive), Indian rock pythons (5 positive), a Burmese python (*P. bivittatus*, 1 positive), green tree pythons (*Morelia viridis*, 2 positive), a carpet python (*M. spilota*, 1 positive), a python of unknown species, and boa constrictors (*Boa constrictor*, 2 positive). The positive samples were from Germany (16), the United Kingdom (7), Denmark (1), the Czech Republic (1), Austria (1), Switzerland (2), France (1), and Belgium (1). Sequencing of the PCR products showed that all were closely related to recently described nidoviruses from pythons with 81-97% nucleotide identity. Virus was mostly detected in oral swabs, although whole blood was positive in 2 cases.

Experimental Ferlavirus Infections in Corn Snakes: Pathologic Alterations and Impact on Lung Exchange Capacity

Michael Pees, Prof, Dr. med. vet., Dipl ECZM (Avian), Dipl ECZM (Herpetology),
Annkatrin Neul, med. Vet.,
Volker Schmidt, Dr. med. vet., Dipl ECZM (Avian) Dipl ECZM (Herpetology),
J. Matthias Starck, Prof, Dr. rer. Nat.,
Rachel E. Marschang, PD, Dr. med. vet., Dipl ECZM (Herpetology)

Session #348

Affiliation: From theClinic for Birds and Reptiles, University of Leipzig, An den Tierkliniken 17, 04103 Leipzig, Germany and the Department of Biology II, LMU München, Großhardener Str. 2, 82152 Planegg-Martinsried, Germany and Laboklin GmbH & Co.KG, Steubenstr. 4, 97688 Bad Kissingen, Germany.

Corn snakes (*Pantherophis guttatus*) were infected with defined ferlaviruses (family Paramyxoviridae) from the genogroups A, B, and C (3 groups with 12 animals each) and pathologic changes were compared to an uninfected control group. Diagnostics included clinical examination, gross pathology, histology and electron microscopy (calculation of exchange capacity of the lungs), as well as virus detection in organs. A scoring system was established to compare the different pathologic alterations among the groups, for the clinical status, lung macroscopy and microscopy, other organ microscopy changes as well as the assessment of lung morphology and the thickness of the air-blood barrier. Relevant microbiologic findings within the lung tissue also were scored. Scoring ranged from 0 (unremarkable) to 2 (severe changes) for each parameter, and a total scoring sum was calculated.

Results demonstrate significant differences in the pathogenicity depending on the virus strain used. This was consistent for all parameters evaluated, resulting in a different pathogenicity scoring between the groups. While the virus strain belonging to genogroup A only caused moderate pathologic alterations, with no snake dying within the experimental period, the isolate belonging to group B, which was originally isolated from a timber rattlesnake, caused the most severe alterations with no snake surviving past day 35 post infection. Lung tissue was most commonly affected, but changes were also noted in the liver, the spleen and the pancreas. Within the lung tissue, a massive thickening of the air-blood barrier was found, accompanied by secondary bacterial infection and inflammatory reactions.

Ferlaviruses and Hemagglutination Inhibition Under Standardized Conditions

Rachel E. Marschang, PD Dr. med. vet., Dipl ECZM (Herpetology),
Annkatrin Neul, TÄ,
Michael Pees, Prof. Dr. med. vet., Dipl ECZM (Avian), Dipl ECZM (Herpetology)

Session #133

Affiliation: From Laboklin GmbH & Co.KG, Steubenstr. 4, 97688 Bad Kissingen (Marschang) and Clinic for Birds and Reptiles, University of Leipzig, An den Tierkliniken 17, 04103 Leipzig (Neul, Pees).

Hemagglutination inhibition assays (HI) are regularly used for the detection of antibodies against ferlaviruses (family Paramyxoviridae) in various species including snakes, lizards, and chelonians. Some studies have shown that there may be extreme differences in HI results between different laboratories. Although some work has been done on serologic comparisons between individual isolates, no data are available on correlation between ferlavirus genogroups and serologic cross-reactivity. Four different ferlavirus genogroups have been described so far (A, B, C, and tortoise). In a previous study, corn snakes (*Pantherophis guttatus*) were inoculated intratracheally with defined ferlaviruses from the genogroups A, B, and C. The animals inoculated with the type B virus developed the most severe disease. Blood was collected from all snakes and from uninfected controls at regular intervals and tested for antibodies against representatives of all 4 known ferlavirus genogroups by HI. Increases in titers were noted 16-28 days post inoculation. The strongest cross-reactivity was noted between the genogroup A and B isolates; the lowest titers were measured against the tortoise isolate. Between the infected groups, the lowest titers against all viruses were noted in the group that received the type B virus, in which the most severe disease was noted. The results of this study show that the development of antibodies against ferlaviruses is influenced by the virus type involved in the infection and that HI results depend on the virus genotype used for testing.

Agamid Adenovirus in Clinically Healthy Bearded Dragons in Italy

Paolo Selleri, DMV, PhD, SpecPACS, Dipl ECZM (Herpetology), Dipl ECZM (Small Mammals),
Nicola Di Girolamo, DMV, MSc (EBHC), PhD, Dipl ECZM (Herpetology),
Alessandro Montani, DMV,
Elia Zanoli, DMV,
Silvia Preziuso, DMV, PhD

Session #274

Affiliation: From Clinica per Animali Esotici, Centro Veterinario Specialistico, Via S. Giovannini 53, 00137, Roma, Italy (Selleri, Di Girolamo, Montani, Zanoli) and Department of Biosciences and Veterinary Medicine, University of Camerino, Via Circonvallazione 93/95, 62024 Matelica (MC), Italy (Preziuso).

Abstract: Adenoviruses infect various species of squamates, chelonians, and crocodiles, with the most common isolate from bearded dragons being the Agamid adenovirus 1 (AAdV-1). Bearded dragons may remain asymptomatic carriers of AAdV-1 or suffer a diverse range of issues, including neurological symptoms, immunosuppression, reduced growth and death. However, true pathogenicity is unclear and Koch's postulates have been only fulfilled for an AdV-induced hepatic necrosis in a boa constrictor *(Boa constrictor).*[1] AAdV-1 has been described in Australia, United States, and Europe.[2] The aim of the present study was to evaluate the presence of AAdV-1 in captive bearded dragons in Italy. For this purpose, 4 private collections were sampled including a total of 39 bearded dragons of various age and sex. Oral and cloacal swabs were obtained from each bearded dragon and were tested by nested PCR to amplify a partial sequence of the adenoviral DNA polymerase gene.[3] Ninety-five percent confidence interval for the prevalence were calculated with the Wilson procedure without a correction for continuity. Out of 39 bearded dragons sampled, 14 resulted positive (prevalence: 36%; 95% CI: 23-52%). Sequencing of the PCR products and comparison with known sequences showed the highest score with AAdV-1. Although results of this study should not be extrapolated in order to obtain the prevalence of AAdV-1 in Italy due to the limited number of samples obtained, these results confirm that AAdV-1 is well established within clinically-healthy captive bearded dragons in private collections in Italy.

References

1. Jacobson ER, Gaskin, JM, Gardiner CH. Adenovirus-like infection in a boa constrictor. *J Am Vet Med Assoc*. 1985;187:1226-1227.

2. Kubiak M. Detection of agamid adenovirus-1 in clinically healthy bearded dragons (*Pogona vitticeps*) in the UK. *Vet Rec*. 2013;72:475.

3. Wellehan JF, Johnson AJ, Harrach B, Benkö M, Pessier AP, Johnson CM, Garner MM, Childress A, Jacobson ER. Detection and analysis of six lizard adenoviruses by consensus primer PCR provides further evidence of a reptilian origin for the atadenoviruses. *J Virol*. 2004;78(23):13366-13369.

Section 26

ARAV Anesthesia and Analgesia

Nicola DiGirolamo, DMV, MSc (EBHC), GPCert(ExAP), PhD, DECZM (Herpetology);
Wallace Stark, III, DVM, MS,
Brad Waffa, MSPH, DVM
Moderators

Snake Anesthesia and Analgesia: A Review

Steve Thompson, DVM, Dipl ABVP (Canine and Feline),
Zdenek Knotek, MVDr, PhD, Dipl ECZM (Herpetology),
Lorraine Corriveau, DVM,
Tomohito Inoue, DVM

Session #113

Affiliation: From Departments of Veterinary Clinical Sciences (Thompson, Inoue) and Veterinary Teaching Hospital (Thompson, Corriveau), College of Veterinary Medicine, Purdue University, 625 Harrison St, West Lafayette, IN 47907, USA, and Avian and Exotic Animal Clinic, Faculty of Veterinary Medicine, University of Veterinary and Pharmaceutical Sciences, Palackeho trida 1946/1, Brno, Czech Republic 612 42, CZ (Knotek).

Advances in anesthetic and analgesic management of lizards[1,2] may not directly benefit snakes despite their close relationship as members of the Order Squamata· Multiple formularies provide doses for practitioners as a starting point without all being based on pharmacokinetic studies and efficacy trials.[3-5]

Ketamine (5-20 mg/kg) is not routinely used alone and is typically combined with lower end doses of diazepam (0.2-0.8 mg/kg) or more commonly midazolam (0.1-2.0 mg/kg). These have been routinely used for sedation and surgical procedures on snakes. Ketamine is not reversible but midazolam can be reversed with flumazenil 0.05 mg/kg. Doses can be IV, IM or SC. Tiletamine and zolazepam (Telazol 100 mg/ml Zoetis, Kalamazoo, MI) has been used at 3-5 mg/kg IM in snakes to facilitate handling and intubation. Doses up to 10 mg/kg may be needed for surgeries. The primary disadvantage of these combinations has been the prolonged recoveries at higher doses necessary for surgical procedures.

Alfaxalone (Alfaxan 10 mg/ml Jurox/Pty Limited, Australia) has been used by one of the authors since 2010 at 1-5 mg/kg IV and provides rapid induction and ability to intubate provided IV access can be obtained. Administration (IM and SC) is effective in reptiles at higher doses of 5-10 mg/kg[6-7] (up to 30 mg/kg reported) and is being explored in combination with opioids. The current challenge with its use relates to label guidelines of single use after opening and cost related to a single individual bottle size.

Propofol (3-10 mg/kg) is also an excellent induction agent provided IV access can be obtained in a larger snake. This also is labeled for single use after opening, which limits extended use for multiple smaller snake patients. Its lipid characteristics result in a higher risk of contamination negating any extended use of a bottle.

General anesthesia with isoflurane or sevoflurane gas can be facilitated in venomous snakes with use of the Pro-Bagger system (Midwest Tongs, Inc, Greenwood, MO). Safety is maintained for the handler and the surgeon during mask induction utilizing a clear plastic handling tube.

Morphine, butorphenol, buprenorphine, hydromorphone and fentanyl are common opioid options available to practitioners. It does not appear that all reptiles share the same affinity for opioid receptors leading to much variability in effect. Butorphenol at 5 mg/kg failed to decrease physiologic stress responses in pythons.[8] Doses of 2 mg/kg had no response in corn snakes[9] but when dosed at 20 mg/kg blunted the response to heat 8 hours after administered. Morphine doses of 10-20 mg/kg that had analgesic properties in bearded dragons did not show benefit in the corn snakes even when increased to 40 mg/kg. Butorphenol doses of 10 mg/kg will produce sedation in pythons but did not provide analgesic effect 7 hours after administration.[10] Buprenorphine at doses of 0.02-1 mg/kg had limited benefit in lizards and turtles[6] with studies on snakes needed. Fentanyl patches (12.5 µg/h) have shown analgesic efficacy in corn snakes and pythons to thermal stimuli.[11] The environmental temperature may affect absorption.

Regional blocks can be used as adjunctive analgesia during surgical procedures. Wellehan and colleagues reported use of mepivacaine at 1 mg/kg in crocodilians.[12] The authors have used mepivacaine at 2-10 mg/kg as a regional block at the site of radiotransmitter implantation in 71 snakes weighing between 400 g and 3 kg.[13] Dilutions may be required in small snakes to use lidocaine (2-5 mg/kg) or bupivacaine (1-2 mg/kg) for local and regional blocks.[14]

Meloxicam has gained routine use with reptiles needing NSAIDs for its anti-inflammatory and pain control properties. Empirical doses appear to be of limited benefit[8] or at too low of a dose. Published[3-5] intramuscular doses begin at 0.1 mg/kg and increase to 0.5 mg/kg in snakes, similar to use in other reptiles. Dose ranges of 0.5-1.0 mg/kg q24h are now routinely used by several of the authors. Hydration and renal status of the patients should be monitored.

References

1. Knotek Z, Standard steps and challenging cases in reptilian anaesthesia. *Proc Annu Conf Southern Eur Vet.* 2011.

2. Knotek Z, Hrda A, Trnkova S. Alfaxalone anaesthesia in green iguana (*Iguana iguana*) *Proc Eur Coll Zool Med 1st Scientific Mtg*. 2011;68.

3. Carpenter JW. *Exotic Animal Formulary*. 4th ed. St. Louis, MO: Elsevier Saunders;2013.

4. Ramsay I. *BSAVA Small Animal Formulary*. 9th ed. Part B: Exotic Pets. Quedgeley, Gloucester: Br Small Anim Vet Assoc.;2015.

5. Plumb DC. *Plumb's Veterinary Drug Handbook*. 8th ed. Ames, IA: Wiley-Blackwell;2015.

6. Sanchez-Migallon Guzman D. Reptile analgesia and anesthesia. *Proc UC Davis Wildl Aquat Anim Med Symp*. 2013.

7. Knotek Z, Hrda A, Knotkova Z, Anestézie želv nádherných intramuskulární aplikací alfaxalonu. (Turtle anesthesia is wonderful with intramuscular alfaxalone.). *Veterinární klinika* 2011;8(2):74-76.

8. Olesen MG, Bertelsen MF, Perry SF, Wang T. Effects of peroperative administration of butorphenol or meloxicam on physiologic responses to surgery in ball pythons. *J Am Vet Med Assoc*. 2008;233(12): 1883-1888.

9. Sladky KK, Kinney ME, Johnson SM. Analgesic efficacy of butorphenol and morphine in bearded dragons and corn snakes. *J Am Vet Med Assoc*. 2008;233(2):267-273.

10. Williams CJA, James LE, Bertelsen MF, Wang T. Tachycardia in response to remote capsaicin injection as a model for nociception in the ball python (*Python regius*). *Vet Anaesth Analg*. 2016.

11. Gutwillig A, Abbott A, Johnson SM. Opioid dependent analgesia in ball pythons (*Python regius*) and corn snakes (*Elaphe guttata*). *Proc Annu Conf Assoc Rept Amph Vet*. 2012:66

12. Wellehan JF, Gunkel CI, Kledzik D, Robertson SA, Heard DJ. Use of a nerve locator to facilitate administration of mandibular nerve blocks in crocodilians. *J Zoo Wildl Med*. 2006;37(3):405-408.

13. Hale V, Thompson S, Corriveau L, MacGowan B, Huse D. Post-surgical Meloxicam use in *Crotalus horridus*. *Proc Annu Conf Assoc Rept Amph Vet*. 2013.

14. Bradley Bay T. Recognizing signs of pain and pain management in exotics. *Proc Conv Can Vet Med Assoc*. 2013.

Comparison and Effect of Anesthetics in Lung Breathing Salamanders *(Ambystoma tigrinum)*

Kerri Cooper-Bailey, MS, DVM,
Stephen Smith, DVM, PhD,
Matthew Close, PhD

Session #152

Affiliation: From Furr, Feathers and Scales 1463 Moore Road, Floyd VA 24091, USA (Cooper-Bailey), Department of Biomedical Sciences, Virginia-Maryland Regional College of Veterinary Medicine (VMRCVM) 205 Duck Pond Dr., Blacksburg, VA 24061, USA (Smith), Department of Biological Sciences, Radford University 174 Center for the Sciences, Radford, VA 24142, USA (Close).

Ambystomatid salamanders are generally robust in size, possess lungs and are capable of cutaneous (skin) respiration and pulmonary (lung) ventilation. As with most salamander species, the cutaneous route of respiration has been the focus of anesthetic studies, research, and medical treatments. However, noting their ability to use other forms of respiration, more avenues should be investigated. Three drugs were used in this study: tricaine mesylate (MS-222), isoflurane gas, and alfaxalone comparing 3 routes of administration injection, inhalation, and immersion. Times for anesthetic induction and recovery, heart rate, respiratory rate, righting reflex, and response to stimuli were recorded before, during and after. An initial preliminary study was performed to determine more accurate doses of alfaxalone since this information is novel for this species. Doses of 10 mg/kg, 15 mg/kg, 20 mg/kg, 25 mg/kg, and 30 mg/kg were chosen in groups of 5 individuals each. The fastest induction periods with the least effect on vitals and shortest recovery were chosen for the second part of the study. Comparison of immersion drugs alfaxalone and MS-222 were compared as well. Alfaxalone at doses of 5 mg/L, and 2000 mg/L were compared to MS-22 at 55 mg/L, 1000 mg/L, 2000 mg/L, with a control. The best results from this group were also moved to the final comparison of all three drugs; MS-222 immersion, inhalation isoflurane, immersion alfaxalone and injectable alfaxalone.

Limiting Transmitter Associated Morbidity and Mortality in Snakes

Steve Thompson, DVM, Dipl ABVP (Canine and Feline)

Session #321

Affiliation: From Departments of Veterinary Clinical Sciences and Veterinary Teaching Hospital. College of Veterinary Medicine, Purdue University, 625 Harrison St, West Lafayette, IN 47907, USA.

Mortality associated with implantation of radio transmitters has been recognized in snakes during conservation studies. Seasonal timing was demonstrated as one of the most significant causes of mortality by Rudolph and colleagues, with a higher risk in smaller snakes.[1] Snakes should not be implanted during the late season when mortality can approach 100%. Lentini and colleagues have reported concerns with radio transmitter morbidity exceeding 66% related to infections and inflammation.[2] This potentially raises concerns with behavior data validity related to decreased or increased movement and basking length and frequency. An Association of Zoos and Aquariums (AZA) task force has expanded specific guidelines for species like the Eastern massassauga rattlesnake (*Sistrurus catenatus catenatus*) advising to not place radio transmitters in snakes under 200 grams, limit transmitter size to less than 50% of snake width and 2.5% of weight in gravid females compared to 5% maximum in males.[3] This also advise to avoid implantation after June in gravid females of this species and after mid-August completely. Expansion to other species can help towards a conservation research goal of documenting post hibernacula survival rates the following spring and lessening any impact on reproduction.

This author seeks to increase dialogue on improving survival and limiting morbidity associated with snake transmitter diseases (STDs).[4,5] Many small transmitters only last a season, so many reports and papers lack any follow up on the snakes' health the following spring. Morbidity and mortality are not routinely published as data in the herpetologic summaries. King and colleagues reported almost 9.5% acute case fatality from anesthesia and post-surgical complications in his small northern population of Lake Erie Watersnakes (*Nerodia sipedon insularum).*[6] Their surgical protocol maintained snakes in captivity for three days prior to release. Debate exists among herpetologists considering time in captivity an additional and unnecessary stressor and favoring immediate release upon recovery.[7] One AZA task force now stipulates that snakes should be kept at least 24 hours after surgery for observation and advises 3-6 days depending on the need for antibiotics considering environmental factors during release.[3] Concerns with a 19% transmitter (Holohil Systems Ltd., Carp, Ontario Canada) failure rate[6] were not observed by this author with their larger transmitters possessing a 2-3 year battery life, allowing multiple years of tracking. No acute mortality was noted in Timber Rattlesnakes (*Crotalus horridus)* with 71 coelomic surgeries and only a single overwintering mortality.[4] Enrofloxacin at 5-10 mg/kg IM q24h × 3 doses (Baytril 22.7 mg/ml. Bayer Healthcare LLC, Shawnee Mission, KS) was provided to all snakes during 48 hours in captivity. Two post-surgical infections were observed yet both snakes survived. One did not have any treatment and the other required transmitter removal and a month of topical and systemic wound therapy for an enrofloxacin resistant strain of Streptococcus.[4] Both snakes were identified alive outside their hibernacula the following spring.

Regional nerve blocks with mepivacaine HCl at 2-10 mg/kg (Carbocaine-V. 20 mg/ml. Pfizer. Pharmacia & Upjohn Company, New York, NY) and injectable meloxicam at 0.2 mg/kg IM q24h × 3 doses (Metacam 5 mg/ml. Boehringer Ingelheim. Fort Dodge, IA) were provided while the snakes were maintained in captivity to assist with pain control and as an anti-inflammatory.[4,5] GIS movement data for 22 snakes that received a replacement transmitter was compared 1 and 2 weeks pre- and post- surgery and found no statistical difference.[5]

ARAV members are encouraged to seek out collaborations and dialogue with herpetologists to improve compliance with IACUC welfare goals regarding surgical technique, asepsis and pain management. This will hopefully lead to reductions in morbidity and mortality and improvement in the conservation data gathered and reported in herpetologic species being studied.

References

1. Rudolph DC, Burgdorf SJ, Schaefer RR, Conner RN, Zappalorti RT. Snake mortality associated with late season radio-transmitter implantation. *Herpetol Rev*. 1998;29(3):155-156.

2. Lentini AM, Crawshaw GJ, Licht LE, McLelland DJ. Pathologic and hematologic responses to surgically implanted transmitters in Eastern massasauga rattlesnakes (*Sistrurus catenatus catenatus*). *J Wildl Dis*. 2011;47(1):107-125.

3. AZA Eastern Massasauga Rattlesnake SSP (2013). Eastern Massasauga Rattlesnake (*Sistrurus catenatus catenatus*) Care Manual. Silver Spring, MD: Association of Zoos and Aquariums;43-44.

4. Thompson S, Corriveau L, MacGowan B, Huse D. Coelomic implantation of Holohil transmitters in *Crotalus horridus*. *Proc Biol Pit Vipers* II. 2014;40.

5. Hale V, Thompson S, Corriveau L, MacGowan B, Huse D. Post surgical Meloxicam use in *Crotalus horridus*. *Proc Biol Pit Vipers* II. 2014;40-41.

6. King RB, Wynn D, Stanford K. Lake Erie Water Snake Annual Report. Available at: http://respectthesnake.com/files/Annual_Report_III.pdf. Accessed September 1, 2013.

7. Ujvari B, Korsos Z. Use of radiotelemetry on snakes: a review. Acta Zool Academ Sci Hung. 2000:46(2):115-146.

Chelonian Anesthesia

Stephen J. Divers, BVetMed, DZooMed, Dipl ACZM, Dipl ECZM (Herpetology), Dipl ECZM (Zoo Health Management), FRCVS, and Scott Stahl, DVM, Dipl ABVP (Avian)

Session #013

Affiliation: From the Department of Small Animal Medicine and Surgery (Zoological Medicine), College of Veterinary Medicine, University of Georgia, 2200 College Station Road, Athens, GA 30602, USA.

Abstract: Chelonians can be a challenge to anesthetize because of their size, demeanor, uncooperative nature, ability to breath hold, and slower metabolism. Following the authors presentation of his approach to chelonian anesthesia, there will be time for questions/answers and discussion on the topic of chelonian anesthesia.

Introduction

There are over 7000 species of extant reptiles exhibiting considerable anatomic, physiologic and pharmacologic differences. The approach to anaesthetizing many pet reptiles is relatively straightforward because they are large enough for intravenous injections and endotracheal intubation, and yet small enough to be easily and safely handled; however, there are occasions when veterinarians are asked to deal with particularly small or large reptiles, and here modifications in standard procedures are often necessary.[1]

One important concept that is often of practical importance is the need to allometric scale drugs doses. For example, while 10 mg/kg propofol may be perfectly appropriate for a 1kg iguana it would probably be a significant overdose for a 75kg komodo dragon. Allometrically scaling the dose, however, provides a much safer and effective dose of 4 mg/kg for the dragon. Similar calculations result in a dose of 20 mg/kg for a 50g gecko.

Pre-Anesthetic Assessment and Stabilization

Fasting should be carried out before all elective surgery in order to reduce compression of lung(s) associated with large meals in carnivores as well as to reduce the chances of regurgitation. In general, 1-2 feeding cycles should be skipped. Fasting has less impact on the volume of hind gut fermenters, including most tortoises. A full history should be taken, assessing not only the animal's current and previous health status, but also paying attention to housing and feeding, since many reptilian ailments relate to sub-optimal environmental conditions or nutritional factors. Accurate identification of species is also essential; this may influence the choice of drugs administered and also the required hospitalization facilities. A full clinical examination should be performed, and the animal accurately weighed. For small reptiles weighing less than 100 grams, an accurate weight to the nearest 0.1 grams is preferred. When dealing with giant chelonians, it may not be possible to obtain an accurate weight or even perform a physical examination until after induction. Weight must therefore be estimated, and the anaesthesiologist should be prepared for under or over-dosing, and serious organ dysfunction that may only become apparent during anesthesia.

The fluid status of all reptiles should be assessed, especially if debilitated or post-hibernation. Determination of packed cell volume and total solids provides a quantitative measure of dehydration, especially if serial samples are analysed. Normal values vary between species, but often lie between 20-35%. The total blood volume of reptiles is around 4-8% of body weight; of this, up to 10% may be withdrawn safely in healthy animals. While inconsequential for large reptiles, blood loss is often a critical factor in small specimens. For example, a 100-gram tortoise that has 0.5 ml of blood drawn for a pre-surgical biochemistry profile and haematology cannot afford to lose any blood during surgery. Therefore, when dealing with very small animals it may be necessary to stage blood collection 7-14 days prior to elective surgery, or forgo pre-surgical blood screens to maintain circulatory volume for non-elective surgery. Giant reptiles may simply be too strong and resist venipuncture even if multiple handlers are employed, or they may represent too great a danger to handle when conscious.

A range of intravenous routes are available for blood collection and the administration of anesthetic and other drugs. The jugular veins are preferred but subcarapacial site (dorsal midline, below cranial carapace), dorsal coccygeal vein (tail), and brachial plexus (deep to the medial tendon of the elbow).

For elective procedures (e.g., neutering), underweight, dehydrated or debilitated animals should be nursed for days, weeks or months until their condition improves. For non-elective surgery attempts to correct dehydration must be started prior to anesthesia. Even the most moribund egg-bound reptile will usually benefit from stabilization for at least 24 hours before embarking on surgery. Whilst oral fluids are non-invasive to administer and provide the most physiologically normal method of rehydration, they are sufficient only for mildly dehydrated animals and are contraindicated immediately prior to surgery due to the risks associated with regurgitation. Intracoelomic fluids are more suitable but uptake can take many hours and their use is obviously problematic if coeliotomy is planned. Thus for dehydrated surgical candidates, intravenous fluid therapy should be administered pre-, peri- and post-operatively as necessary. Fluids are administered at 20-35 ml/kg/day (rates of up to 5 ml/kg/hour may be initially used for the first 3 hours); the use of an infusion pump or syringe driver achieves accurate dosing but multiple bolus injections are a practical alternative. The placement of an indwelling jugular catheter, or better still, a long stay central line can greatly facilitate vascular access.

All reptiles should be hospitalized and maintained within their preferred optimum temperature zone to minimize physiologic disturbance and facilitate recovery and immunocompetence. While small animals can be easily housed in dedicated hospital vivaria, maintaining a 100 kg giant tortoise at 28°C often presents practical difficulty. Hypothermia also affects the pharmacokinetics of administered drugs and greatly prolongs recovery. When dealing with large reptiles it is often necessary to plan ahead. It may be necessary to acquire additional personnel for lifting and moving the animal, and careful consideration needs to be given to access and movement around the hospital.

The calculation and print out of an emergency drug sheet is recommended (Table 1). Such documents are best prepared using EXCEL so that drug calculations are automatic, and they can simply be printed and attached to the patient's medical record.

Table 1. Reptile anesthesia and emergency drug calculator.

Insert reptile weight, print, and attached to anesthesia form.					
Patient name:	Tortie				
Patient number:	307892				
Patient weight (kg):	2.671				
Drug	Concentration (mg/ml)	Dose (mg/kg)	Route	Bolus	Units
Atropine	0.54	0.04	IV	0.20	ml
Epinephrine (1:1000)	1	0.1	IV	0.27	ml
Calcium gluconate	100	100	IV	2.67	ml
Furosemide	10	2	IV	0.53	ml
Maintenance fluids (ml/kg/day)		7.5	IV	0.83	ml/h
Intraoperative fluids (ml/kg/h)		3	IV, IO	8.01	ml/h
Shock fluid bolus (ml/kg/h)		10	IV, IO	27	ml/h
Hetastarch 6%		5	IV, IO	13	ml/h
Morphine	1	1.5	IM	4.01	ml
Hydromorphone	2	0.5	IM	0.67	ml
Propofol	10	10	IV	2.67	ml
Midazolam	5	2	IM	1.07	ml
Meloxicam	5	0.2	IM	0.11	ml
Lidocaine	20	2	Local	0.27	ml

Premedication and Analgesia

Premedication using midazolam may be useful, especially if combined with an opiate. Recent studies have shown that intramuscular injections can take an hour or more to reach peak levels. The author frequently uses intramuscular midazolam (1-2 mg/kg) and hydromorphone (0.5 mg/kg) for chelonians. There is still controversy concerning the use of opiate analgesics in reptiles, and taxa-specific effects have been documented; however, morphine, hydromorphone and tramadol appear effective for chelonians.[2-4] Meloxicam (0.2 mg/kg) is used routinely for all species, and may be repeated after 48 - 72 hours if necessary.

Induction of Anesthesia

Propofol, given intravenously at doses of 3-15 mg/kg depending upon size provides a rapid, controlled method of induction. It is relatively non-toxic and there is reduced risk of thrombophlebitis if injected perivascularly; this is of particular concern since intravenous access may be relatively difficult, especially in active animals undergoing elective procedures. Alfaxalone (also known as alphalaxone) is also FDA approved and is typically used at similar IV dose rates of 3-12 mg/kg; however, unlike propofol, alfaxalone can also be effective at higher IM doses of around 10-20 mg/kg (Table 2). These IM dose volumes are high and should be divided into several injection sites. Unfortunately, alfaxalone must be used or discarded within 6 hours which can make it cost-prohibitive.

The use of ketamine and (dex) medetomidine provides a reversible light to surgical plane of anesthesia in many chelonians. For intramuscular injections, the forelimb musculature is preferable. For small reptiles, accessing a muscle mass can be difficult, in which case subcutaneous administration may be the only option (although there does not appear to be a significant difference between induction times in most cases).

Table 2. Anesthesia drugs, dosages, and routes of administration.

Drug/Dosage	Route	Comments
Propofol (3-12 mg/kg)	IV	Dependent on intravascular administration; poor response often associated with partial perivascular injection.
Alfaxalone		Must be discarded after 6 hours.
(3-12 mg/kg)	IV	Similar results to propofol.
(10-20 mg/kg)	IM	Large volume must be divided into two or more injections.
Dexmedetomidine (0.025-0.075 mg/kg) plus Ketamine (5-25 mg/kg)	IM	Can produce inconsistent results in some species, especially terrestrial tortoises, long recoveries unless reversed using atipamezole; preferred for small chelonians where IV access and intubation are not possible.
Telazol (3-10 mg/kg)	IM	Prolonged recovery at higher doses.

Intubation

Chelonians, all aquatic and large terrestrial species, can have powerful bites, and a mouth gag should always be used. A stylet within the endotracheal tube also greatly facilitates placement. Airway resistance becomes increasing critical as endotracheal tube diameter decreases, because resistance, (Table 3).

Table 3. Relative airway resistance within endotracheal tubes from 1.0-4.5 mm in internal diameter.

Endotracheal tube diameter (mm)	Relative resistance
1.0	1000
1.5	198
2.0	63
2.5	26
3.0	12
3.5	7
4.0	4
4.5	2

Maintenance of Anesthesia

Isoflurane or sevoflurane are the agents of choice for maintenance of anesthesia. These volatile inhalation agents have quicker induction and recovery times than most gaseous alternatives. While the differentials in cost, and faster induction/recovery associated with sevoflurane, compared to isoflurane, are insignificant for small reptiles, there are increasingly significant for giant reptiles. Furthermore, the lack of reliance on hepatic metabolism or renal excretion further reduces the anesthetic risk to debilitated reptiles or those with questionable renal or hepatic function; however, both isoflurane and sevoflurane cause dose-dependent hypotension.

Following induction and intubation, chelonians, especially large animals, should be maintained in sternal until a surgical plane is achieved. Placing a large chelonian into lateral or dorsal recumbency can result in ventilation-perfusion mismatch.

Non-crocodilian reptiles lack a muscular diaphragm, instead relying on active gulps of air mediated by the action of skeletal intercostal muscles (squamata, crocodilia) or limb movements (chelonia). The action of these muscles is generally abolished at surgical planes of anesthesia; intermittent positive pressure ventilation (IPPV) should therefore be used routinely and will also help prevent small endotracheal tubes from becoming blocked. Ventilation every 10-120 seconds should typically be used and modified to maintain $ETCO_2$ levels of between 15-35 mmHg. Ideally, a pressure-cycle (e.g., BAS Vetronics, Bioanalytical Systems Inc, West Lafayette, IN, USA) for small reptiles, or a volume-cycle ventilator (e.g., Veterinary Anesthesia Ventilator 20021E, Hallowell Inc, Pittsfield, MA, USA) for large reptiles should be used. End-tidal CO_2 values <10mmHg tend to result in prolonged apnea and recovery.

For chelonians that are too small to intubate, the author relies on injectable agents (often ketamine-dexmedetomidine) and combination with local anesthesia (e.g., lidocaine).

Peri-operative fluid support, temperature control, and monitoring are important. Warm water circulating pads and forced warm air blankets are preferred to maintain the core body temperature within the species-specific preferred temperature range.

Monitoring During Anesthesia

A dedicated anesthesia recording form should be used routinely (Table 4). Palpebral and corneal reflexes are reliable in those species in which they can be elicited; however, corneal reflexes are abolished at toxic doses, and pupillary diameter bears little relation to the depth of anesthesia (unless fixed and dilated which indicates excessive anesthetic depth or brain anoxia and death); however, jaw tone (being careful not to get bitten!) and withdrawal reflexes (tongue, limb or tail) are useful, becoming abolished only at a surgical plane of anesthesia. This also correlates with full loss of righting reflex, loss of spontaneous movement and complete muscle relaxation. Cloacal reflex and tone is generally only abolished at excessive anesthetic depths.

Heart rate is often most reliably documented by doppler. Pulse oximetry, with an oesophageal or cloacal probe, is useful in monitoring pulse rate and strength. Although the spO_2 readings often under-estimate saO_2 values in most reptiles, monitoring the pulse oximetry trend may be more useful than the absolute values. Doppler can also be very effective and used over peripheral arteries or the heart; however, indirect blood pressure is inaccurate and untrustworthy compared to direct measures, and cannot be relied upon. $ETC0_2$ is especially useful to ensure appropriate ventilation. Using controlled ventilation and accurate cardiovascular monitoring it is possible to accurately control anesthesia. Arterial samples for blood gas evaluations are often difficult to obtain. Nevertheless, pH, PCO_2 and PO_2 are measured directly and can be trusted. Other values are calculated using human algorithms. Arterial samples are required for PO_2 and SaO_2 evaluation, but venous samples can be useful for other parameters including pH, lactate, bicarbonate, TCO_2, and BE.

Towards the end of surgery, the anaesthetic gas should be stopped while maintaining IPPV to facilitate inhalant drug excretion. Those under the influence of injectable agents should have them reversed (e.g., atipamezole for dexmedetomidine). Subsequently, ventilation should be reduced to every 1-5 minutes using room air to encourage mild hypoxia and hypercapnia, increase respiratory drive, and initiate a return to spontaneous breathing. It is also helpful to flex, extend and massage the limbs to encourage venous return and improved circulation. The administration of epinephrine at 0.1 mg/kg IM appears to reduce recovery periods significantly.

Table 4. Reptile anesthesia record.

Abbreviations									
Time	0	15	30	45	60	15	30	45	60
O_2 flow L/min									
Vaporizer %									
Inspiratory gas %									
End tidal gas %									
Inspiratory O_2									
End-tidal CO_2									
SpO_2									

HR ●

Ventilation ○
S Spontaneous
C Controlled

Arterial pressure
▽ Systolic
X Mean
△ Diastolic

Left scale	Right scale
85	400
80	380
75	360
70	340
65	320
60	300
55	280
50	260
45	240
40	220
35	200
30	180
25	160
20	140
15	120
10	100
5	80
0	0

Temperature

CRMM

Anes induct____________.

Gas started____________

Surgery start___________.

Surgery end____________.

Anes end______________.

Extubation_____________.

RECOVERY (in cage)

RR Respiratory rate
SM Spont. Movement ☑

Time q____ mins		
Time	RR	SM

Comments:

ET tube size:____.

Recovery and Post-Operative Support

Once the reptile is breathing spontaneously it can be returned to the hospital enclosure to fully recover. It is impractical to maintain the animal on the operating table until completely recovered as this may take many hours when giant reptiles are involved. Nevertheless, trained technicians should continue to monitor the reptile every 5-15 minutes until spontaneous movement and righting reflexes return. Only then is safe to leave the animal unsupervised. Aquatic species should not be returned to water for 12-24 hours. Oral and injectable tramadol has been shown to provide analgesia with less respiratory depression in chelonians, and should be considered for continued post-operative analgesia.[5]

References

1. Schumacher J, Mans C. Anesthesia. In: Mader DR, Divers SJ, eds. *Current Therapy in Reptile Medicine and Surgery*. St Louis, MO: Elsvier;2014:134-153.

2. Sladky KK. Analgesia. In: Mader DR, Divers SJ, eds. *Current Therapy in Reptile Medicine and Surgery*. St Louis, MO: Elsvier;2014:217-228.

3. Baker BB, Sladky KK, Johnson SM. Evaluation of the analgesic effects of oral and subcutaneous tramadol administration in red-eared slider turtles. *J Am Vet Med Assoc*. 2011;238(2):220-227.

4. Sladky KK, Miletic V, Paul-Murphy J, Kinney ME, Dallwig RK, Johnson SM. Analgesic efficacy and respiratory effects of butorphanol and morphine in turtles. *J Am Vet Med Assoc*. 2007;230:1356-1362.

5. Baker BB, Sladky KK, Johnson SM. Evaluation of the analgesic effects of oral and subcutaneous tramadol administration in red-eared slider turtles. *J Am Vet Med Assoc*. 2011;238(2):220-227.

ARAV

ARAV Master Classes and Roundtables

David Hannon, DVM, DABVP (Avian);
Kathleen Marie Heinz-Taheny, DVM, PhD, Dipl ACVP;
Rachel Marschang, PhD, Dr med vet, DECZM (Herpetology), FTÄ Mikrobiologie, ZB Reptilien
Moderators

Gross Lesion Recognition for the Reptile Clinician

Michael M. Garner, DVM, Dipl ACVP

Session #008

Affiliation: From Northwest ZooPath, 654 West Main, Monroe, WA 98272, USA.

Reptile veterinarians in clinical practice, by nature of their specialty, often have to select diagnostic procedures that will provide the best and least expensive treatment options. A familiar course is establishing a viable differential diagnostic list during clinical exam or necropsy based on the gross appearance of the lesions. Gross lesion recognition (and photography), like other imaging modalities, is a bit of an art form. Vast zones of grey may confront the investigator when considering what the lesion is, what it could be, and what to do with it. And nobody is good at all of it: A seasoned livestock pathologist could possibly clear an entire days' necropsies without the help of a single histology slide. A bee keeper would recognize immediately the maggot stages that parasitize his colony. A dog-cat pathologist in a busy private laboratory or a companion animal clinician may not have a clue about gross lesions in livestock or bees. Historically, our attempts at accurately diagnosing disease by gross lesion recognition have been a humbling experience, and that has been the impetus for all further diagnostic specialties. Gross lesions, or lack thereof, are the first visual indication of what may be wrong with the patient (or cadaver, as the case may be). These are the lumps, the effusions, the asymmetrical oddities, the discolorations, the odiferous clues to disease that (hopefully) stimulate a "scientific" thought process culminating in a list of differential diagnoses. The purpose of this 2-hour interactive workshop is to present clear and not-so-clear images of common and not-so-common gross lesions in live and necropsy specimens, and in a participatory manner, establish a differential diagnosis and means for establishing a definitive diagnosis.

Recommended Bibliography

1. King JM, Dodd DC, Roth L, Newson ME. *The Necropsy Book*. Charles Louis Davis Foundation; 2002. (Mammals)

2. Reimschuessel R. Postmortem examination. In: Stoskopf MK, ed. *Fish Medicine*. Philadelphia, PA: W.B. Saunders:160-165. (Fish)

3. Garner, MM. Overview of biopsy and necropsy techniques. In: Mader DR, ed. *Reptile Medicine and Surgery*. 2nd ed. St. Louis, MO: Elsevier Science;2006:569-580. (Reptiles)

4. Nichols DK. Necropsy. In: Wright KM, Whitaker BR, eds. *Amphibian Medicine and Captive Husbandry*. Malabar, FL: Krieger Publishing Co.;2001:331-332. (Amphibians)

5. Latimer KS, Rakich PM. Necropsy examination. In: Ritchie BW, Harrison GJ, Harrison LR, eds. *Avian Medicine: Principals and Application*. Lake Worth, FL: Wingers Publishing, Inc.;1994:356-376. (Birds)

6. Campbell TW, Ellis C. *Avian and Exotic Animal Hematology and Cytology*. Ames, IA: Iowa State University Press;2007.

In-Depth Review of the Cranial Coelom (Thorax) of Reptiles: Normal Anatomy

Jeanette Wyneken, PhD

Session #349

Affiliation: From Department of Biological Sciences, Florida Atlantic University, Boca Raton, FL 33431, USA.

The reptilian thorax is defined here as the cranial coelomic cavity and its surrounding skeleton. Major structures are located cranial to the duodenum. Because no species has a diaphragm partitioning the pleuoperitonal cavity, the designation used here is based on structural and functional relationships. The esophagus, stomach, trachea, bronchi and lungs, liver, heart and great vessels, as well as major supporting mesenteries, occupy the cranial coelomic cavity. For the purposes of this review, the liver is discussed superficially because its major functions align more closely with caudal coelomic structures.

Skeleton

The trunk vertebrae, ribs (or cranial carapace), sternum and interclavical (or cranial plastron) and their muscles surround the thorax. In squamates (snakes and lizards), crocodilians and tuataras, the cranial aspects of the coelom are bounded by the trunk vertebrae (sometimes termed thoracic vertebrae) and ribs. Chelonians (turtles and tortoises) vary greatly as to how far the caudally the cranial organs extend dorsally so carapacial landmarks tend to be less useful than plastral landmarks. The cranial coelom can extend to the hyoplastron bones (usually demarked by the humeral and abdominal scutes). Normal trunk vertebrae in lizards, snakes and crocodilians have procoelous joints articulating vertebral bodies. Procoelous vertebrae each have cavity on the cranial aspect and a convex surface on the caudal end.[1,2] In all turtles and tortoises, the trunk vertebrae, ribs and ventral ribs (gastralia, abdominal ribs) are modified to form the shell's carapace and plastron.[3,4] The trunk vertebrae are neural bones and these overlie vertebral bodies, each with amphicoelous articulating ends. The ribs are flattened and expanded as pleural bones. In tuataras, the vertebrae are amphicoelous (having cavities at both ends), similar to those of fish.[5]

Many lizards, crocodilians and tuataras have ribs associated with the trunk vertebrae, with more cranial ribs joining the interclavicle and sternum via cartilaginous joints. Most posterior ribs are floating. These taxa also have ventral ribs that can extend sufficiently far cranially to reach the cranial coelom. All snakes lack an interclavicle, sternum and gastralia. Chelonian ribs are modified as flattened pleural bones and that articulate with vertebrae forming the carapace.

Tuataras are distinctive among reptiles in having short uncinate processes arising roughly orthogonally along the caudal edge of each rib. Uncinate processes are similar to those found in birds and some dinosaurs.[1] Tuataras also have floating ribs that do not attach to the sternum.

Divisions of the Coelom

The thorax is partitioned into the pleuroperitoneal cavity and pericardial cavity. The latter contains the heart and the roots of the heart with the roots of the great vessels. Both are surrounded by the pericardial sac, which is part of the transverse septum. The pleuroperitoneal cavity extends to the pericardial sac ventrally; it extends more cranially dorsal to the pericardium (Fig 1). The pleuroperitoneal cavity, surrounded by the peritoneum (coelomic membrane) contains the other visceral organs except for the kidneys, which are retroperitoneal. Pleural cavities form secondarily in chelonians, crocodilians and some lizards via a nonmuscular, thin fold of tissue that grows from the median mesentery supporting the gut to the peritoneum.[6]

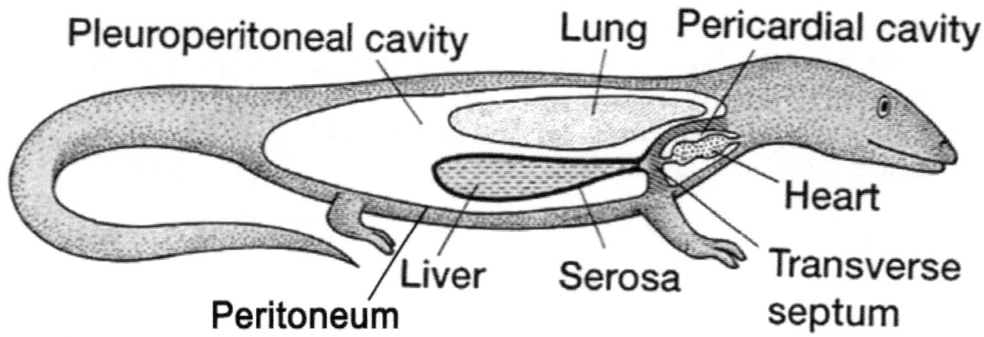

Figure 1. Body cavities of a generalized reptile. The serosa of the liver is formed from the posterior wall of the transverse septum (modified from Kardong, 2014 with permission).

The liver fills much of the cranial aspects of the pleuroperitoneal cavity, ventral to the lungs, with left and right lobes that may or may not join one another via a bridge of hepatic tissue. The liver attaches to the caudal pericardium via a short coronary ligament (homologous to the falciform ligament of mammals). The paired ventral abdominal veins (in chelonians) or single abdominal vein (squamates, crocodilian and *Sphenodo*n) travels through the peritoneum and passes into the liver through the coronary ligament.

Lungs, Trachea and Bronchi

The trachea, caudal to the larynx, passes through the thoracic inlet and soon bifurcates into right and left extrapulmonary bronchi in chelonians, lizards, crocodiles and *Sphenodon*. Snake trachea and bronchi may be extrapulmonary or pulmonary (within the lung) depending on species. For example, crotalids (rattlesnakes) have a single tracheal lung whereas boids (boas and pythons) have paired short extrapulmonary bronchi. In lizards and *Sphenodon*, no intrapulmonary bronchi are present and the extra-pulmonary bronchi, when present, tend to be short. Chelonians have bronchi that extend well into the lung in most species. Crocodilians have well-developed branched bronchi that direct airflow unidirectionally.[7] The trachea is formed of c-shaped cartilaginous rings. Bifurcation into bronchi is distal and initially hidden (grossly) within the tracheal tube. The lining of the trachea and bronchi is formed of ciliated, secretory and basal epithelium.[8] The lungs lie dorso-caudal to the heart and dorsal to the cranial parts of liver and stomach except on snakes.

Esophagus and Stomach

The esophagus passes into the cranial thorax centrally and as it passes caudally may lie dorsal to the trachea or slightly to one side. In cheloniid sea turtles and some freshwater turtles, the esophagus takes an s-shaped bend to the left before joining the stomach. The length of the esophagus is often related to the neck length except in snakes, legless lizards, and leatherback sea turtles (*Dermochelys coriacea*). It is particularly long in monitor lizards, snakes and some turtles (such as softshelled turtles, long-necked Australian pleurodires and mata matas [*Chelys fimbriata*]). In the leatherback sea turtle, the esophagus is particularly long, extending to the pelvic area then turning into the left and travelling cranially to the level of the axilla where it then turns medially and joins the stomach near the heart. The lining of the esophagus is smooth or folded and usually lined by stratified squamous epithelium with ciliated cells and goblet cells,[8] except in sea turtles that have large keratinous papillae lining the esophagus. These reduce dramatically in size at the gastroesophageal junction. The gastro-esophageal junction has a sphincter. The esophagus is primarily a transport organ. However, the mucosal wall contains lymphoid tissue patches, termed tonsils (gut-associated lymphoid tissue) that are particularly obvious in boid snakes.[8] The esophagus is very thin in snakes, particularly cranially. It has more pronounced muscular layers in tortoises and sea turtles.[8,9] Distally, in all reptiles, the esophagus has muscular layers near the stomach.

The stomach and esophagus tend to join to the left of the heart and often along the dorsal surface of the liver's left lobe. In all reptiles, the stomach is located on the left side of the body and attaches to the left lung (when present) by a gastropulmonary ligament. In snakes, the stomach is fusiform whereas in most lizards and turtles it is sac-like and c-shaped. The stomach of crocodilians has two parts: a large fundic portion that receives the esophagus and a smaller pyloric portion that connects proximally to the fundic part by a large orifice and distally to the duodenum.

The stomach is primarily a chemical digestion organ, except in crocodilians that also have a muscular pyloric region that functions in physical digestion. In snakes, the stomach is rather short and the gastroesophageal junction is located at approximately the caudal one-quarter point of the body and along the left side of the liver. In chelonians, the stomach is sac-like, usually c-shaped, and usually lies along the heart's left side and curves to the right. The stomach ends in a muscular pyloric sphincter, the pyloroduodenal junction.

The lining of the stomach is glandular and its surface mucosa can be very dynamic in shape and thickness.[9] In episodic feeders such as many snakes and some crocodilians, the lining of the stomach may be pale and folded during periods of fasting. After feeding, the mucosa of the stomach and intestine become pink and the epithelial cells elongate, thickening the stomach's mucosa and increasing area for absorption. When full, the stomach's muscular wall can be stretched thin.

Heart and Major Vessels

The heart is located medially in the pericardial cavity, at the level of the shoulder girdle in most lizards, just caudal to the shoulders in chelonians, approximately midway between the bases of the fore- and hindlimbs in crocodilians and with the second quarter of the length of snakes, caudal to the head.[10] The heart is connected caudally via the ascending vena cava to the liver and cranially via the external jugular veins from the head and neck.

The hearts of turtles, lizards, and snakes are grossly similar. In ventral view, the heart appears to be a three-chambered pump composed of two muscular atria and one ventricle; however, dorsally the large, thin-walled sinus venosus is found which is part of the heart.[10-12] The sinus venosus receives venous blood returning from the head, limbs, and body that then flows to the right atrium. Three great vessels emerge from the cranioventral aspect of the ventricle. From left to right, these are the left and right aorta and the pulmonary trunk. Crocodilians have a heart structure that differs from the other reptiles. The ventricle is divided into pulmonary and systemic sides. The left aorta and pulmonary artery arise from the right ventricle; the right aorta drains the left ventricle. Crocodilians have a connection near the bases of the two aortae, the Foramen of Panizza, which appears to function like a central shunt.[6]

The major vessels passing through the cranial coelom include (but are not limited to) the left and right aortae, pulmonary arteries and veins, celiac artery, subclavian arteries and veins, proximal parts of the carotid arteries, post cava, azygos vein, caudal parts of the external jugular veins, hepatic veins, and cranial part of the hepatic portal vein.

Thyroid, Thymus, Parathyroid and Ultimobranchial Glands

The cranial coelomic cavity houses several endocrine glands. All develop as diverticula from the pharyngeal folds. The thyroid gland is a single structure in snakes (round or oval) and lizards (U-shaped) and is located ventral and lateral to the trachea and cranial to the heart; near the brachiocephalic trunk artery in chelonians (round gland). In crocodilians, glands are found along the ventrolateral trachea near the bronchi as a bilobed gland or as or two separate lobes. In Nile crocodiles (*Crocodylus niloticus*) they are found along the bronchii depending on species.[6,13]

Thymus glands are lobulated glands that are paired in chelonians and squamates, and usually single in crocodilians. In turtles they are associated with the common carotid and subclavian arteries.[8,10] In lizards and snakes, the thymus glands are associated with the common carotid arteries, jugular veins and vagus nerve.[8] In crocodilians, the main body of the thymus is a singular structure just cranial to the heart and has a cranial extremity that extends to the cervical region near the base of the skull.[8]

The parathyroid glands and ultimobranchial bodies are small, antagonistic glands that regulate blood calcium. There are one, two or three pairs of parathyroid glands (raises blood calcium) in reptiles and, usually, a single pair of ultimobranchial bodies (lowers blood calcium).[6] They are usually associated with the carotid arteries. In chelonians and crocodilians one of two pair(s) can be found near the heart. In turtles and tortoises another pair(s) may be associated with the thymus and sometimes may be palpated as dense bodies along the thymus' connective tissue.[8] In lizards, the pairs of parathyroids tend to be associated with the carotid arties near the heart. In snakes, these glands are often distributed along the carotid arteries, distal to the thymus.[8]

The ultimobranchial bodies are difficult to find and are usually near the thyroid gland. Jacobson (2007) notes that many lizards, crocodilians, and snakes have just one ultimobranchial body on the left side; the right does not persist.[8] Chelonians and, at least sand boas (*Eryx johnii*), have both right and left glands.[8]

References

1. Romer AS. The axial skeleton. In: *Osteology of the Reptiles*. Chicago, IL: University of Chicago Press;1956.

2. Molnar JL, Pierce SE, Bhullar B-AS, Turner AH, Hutchinson JR. Morphological and functional changes in the vertebral column with increasing aquatic adaptation in crocodylomorphs. *R. Soc Open Sci.* 2015;2(11):150439:1-22. http://dx.doi.org/10.1098/rsos.150439.

3. Pritchard PCH. Evolution and structure of the turtle shell. In: Wyneken J, Godfrey MH, Bels V, eds. *Biology of Turtles: From Structures to Strategies of Life*. Boca Raton, FL: CRC Press;2008:45-84.

4. Rice R, Kallonen A, Cebra-Thomas J, Gilbert SF. Development of the turtle plastron, the order-defining skeletal structure. *Proc Natl Acad Sci* USA. 2016;113(19):5317. doi:10.1073/pnas.1600958113

5. Gadow H. *Amphibia and Reptiles*. London: Macmillan and Co. Limited;1901.

6. Kardong, KV. *Vertebrates: Comparative Anatomy, Function, and Evolution*, 7th ed. New York, NY: McGraw-Hill;2014.

7. Farmer C, Sanders K. Unidirectional airflow in the lungs of alligators. *Science*. 2010;327(5963): 338-340.

8. Jacobson, ER. *Infectious Diseases and Pathology of Reptiles: Color Atlas and Text*. Boca Raton, FL: CRC Press Taylor & Francis Group;2007.

9. Parsons, TS, Cameron JE. Internal relief of the digestive tract. In: Gans C, Parsons, TS, eds. *Biology of the Reptilia*. Vol 6. Morphology E. New York, NY: Academic Press;1977:159-223.

10. Wyneken. J. *Guide to the Anatomy of Sea Turtles*. NMFS Tech Publication. 2001;NOAA Tech Memo NMFS-SEFSC-470.

11. Wyneken J. The structure of cardiopulmonary systems of turtles. In: Wyneken, J, Godfrey M, Bels V, eds. *The Biology of Turtles*. Boca Raton, FL: CRC Press Taylor & Frances Group;2008:213-224.

12. Wyneken J. Normal reptile heart morphology and function. *Vet Clin North Am Exot Anim Pract*. 2009;12(1):837-853.

13. Huchzermeyer FW. *Crocodiles: Biology, Husbandry and Diseases*. Cambridge, MA: CABI;2003.

In-Depth Review of the Cranial Coelom (Thorax) of Reptiles: Disease Conditions

Drury R. Reavill, DVM, Dipl ABVP (Avian), Dipl ABVP (Reptile and Amphibian), Dipl ACVP

Session #199

Affiliation: From Zoo/Exotic Pathology Service, 6020 Rutland Drive #14, Carmichael, CA 95608, USA

Abstract: This paper will cover the disease conditions of the reptilian cranial coelom (thorax). The organs include the thyroid, thymus, heart, trachea, lungs, esophagus, and stomach. Common abnormalities in these organs include developmental anomalies, degenerative lesions, obstructions, traumatic injuries, infectious disease agents, and neoplasms. The paper is divided by disease categories and location. A short discussion of these disease conditions in reptiles is found in each section. For the purposes of this paper, chelonia will refer to all of the shelled reptiles. The term tortoise will be used to describe the large land animals that have stout elephantine feet. Few references to marine turtles and crocodilian species are included, as these are uncommonly seen in most clinical practices.

Non-Infectious Diseases

Developmental anomaly

Developmental anomalies of the cardiovascular system are reported to be common in reptiles[1] and are usually associated with deaths of juveniles.[2] Incomplete development of the atrioventricular valves has been described, leading to incomplete valve closure.[3] A secundum atrial septal defect was described in a Komodo dragon (*Varanus komodoensis*).[4] The lesion was on the craniodorsal portion of the septum. It led to chronic congestion of several organs. Two juvenile ball pythons (*Python regius*) with bifid ventricles were reported.[5] All cardiac chambers were enlarged, the lesion disrupted the ability to separate pressures within the ventricles. An interventricular septal defect was also reported in an alligator (*Alligator mississipiensis*).[6]

A number of congenital disorders of the digestive tract and rarely the respiratory tract are recognized in reptiles even from the early Cretaceous age (dicephaly).[7] Ophidian and chelonian dicephaly (craniodichotomy) is presumably the result in incomplete separation of identical twins and is presumed to be the cause in other reptiles.[8,9] Rarely dicephaly has even been used in experimental studies on causes of satiety.[8] Tracheal duplication and malformation has been described in an adult (7-year-old) bicephalic Honduran milk snake (*Lampropeltis hondurensis*). This snake also had two esophagi and two stomachs that joined at a common duodenum.[10]

In a spur-thighed tortoise (*Testudo graeca ibera*) the two heads and necks were independent, with duplicated esophagus, stomach, liver, gallbladder, trachea, and two asynchronous hearts.[9] The proposed causes reported are exposure to environmental toxins, genetic defects, anoxia of the embryo, elevated environmental temperatures during incubation, and rapid thermal changes during internal incubation.[8,9]

Degenerative

An incidental finding of oxalate crystals in thyroid colloid has been described in wild desert tortoises, (*Gopherus agassizi*). The crystals are birefringent radiating aggregates and not associated with inflammation as those identified in some of the tortoise kidneys. It was felt the origin of the calcium oxalates was from the plants eaten.[11]

Few diseases are reported involving the thymus of reptiles. Thymic hyperplasia in subadult captive-bred Galapagos tortoises (*Geochelone nigra*) resulted in significant cervical swelling near the thoracic inlet.[12] Although thymic enlargement can occur normally in reptiles as a result of seasonal hormonal changes, these were felt to be excessive given the mass lesion.

Aortic aneurysm rupture leading to cardiopulmonary arrest was described in a Burmese python (*Python molurus bivittatus*) and the exact cause was not determined.[13] Arteriosclerosis has been documented in green iguanas (*Iguana iguana*)[14] and dystrophic mineralization of arteries is seen in various lizards. Mineralization of arteries is most likely due to secondary nutritional hyperparathyroidism due to a calcium deficiency. A mole king snake (*Lampropeltis calligaster rhombomaculata*)[15] and a Deckert's rat snake (*Elaphe obsoleta deckertii*)[16] were diagnosed with cardiomyopathy of undetermined cause. The lesions in the king snake were primarily collagen proliferation and osteoid-like material, while the rat snake had degeneration and necrosis of myocardial fibers. Aortic stenosis with associated dilatation of the right atrium and the ventricle were found in an iguana.[17] Histologically both aortic arches had narrow lumens with thickened intimas, and the myocardial fibers were attenuated. Although the cause was not determined the authors considered a chronic congenital lesion as one possibility.

Myocardial degeneration due to vitamin E deficiency has been seen in several reptile species.[18] Grossly and microscopically the lesion is similar to that described in mammals. The heart will have mottled white to grey myocardium and histologically there will be multifocal to coalescing loss of myocytes replaced with dense fibrous stroma. Visceral gout can lead to severe pericardial lesions,[19] as well as changes in the myocardium. These can include thickening of the pericardial sac with a deposition of the urate crystals as well as crystals forming within the myocardium. Primary disease can be nutritional or toxic.[20]

It has been reported that the most common noninfectious cause of respiratory disease in chelonians results from trauma to the carapace.[21] Though not well documented in the literature, pulmonary mineralization is occasionally seen secondary to osteodystrophy from any cause. In some cases, this is limited to pulmonary blood vessels, but basement membranes may also be affected. Pleural urate deposition can be seen in cases of visceral gout.[22] Inhalation of foreign material can lead to serious lung lesions and predispose to bacterial infections.[23]

Obstructive lesions

The most common cause of gastrointestinal obstructions is from the ingestion of foreign material.[24,25] Generally the foreign material is of inappropriate environmental substrates.

A variety of foreign objects have resulted in obstructive lesions in the digestive tract of snakes.[24,25] A black rat snake (*Elaphe obsoleta obsolete*) ingested a golf ball which lodged within the stomach.[26] It was successfully removed surgically and the snake returned to the wild.

An unusual case of parasitic impaction was reported in an adult make milk snake (*Lampropeltis triangulum*).[27] The snake was anorexic for two months and had regurgitated a partially digested mouse. Several fluid-filled coelomic masses were identified by radiographs and ultrasound. A gastric impaction was due to a firm solid white mass containing the eggs of the nematode *Kalicephalus* sp and the coelomic masses were granulomatous inflammatory foci secondary to the nematode eggs.

A review of post-mortem examinations on three free-living green turtles (*Chelonia mydas*) in the United Arab Emirates identified a duodenal volvulus and stomachs full of fresh seagrass. Based on a lack of other findings, the duodenal volvulus was suspected to be dietary in origin from over-fermentation of ingesta.[28]

Gastric impaction due to food material has been described in captive crocodiles (*Crocodilus niloticus*).[29] Five crocodiles fed only guinea pigs for 4 days died and in the stomach was a compacted mass of matted guinea pig hair.

Infectious Diseases

Heart

Infectious agents affecting the heart are usually secondary to systemic illness.[30] West Nile virus infection led to myocardial degeneration and necrosis in a farmed American alligator (*Alligator mississippiensis*).[31]

Bacterial disease has been reported in conjunction with endocardial thrombosis,[32] endocarditis,[33] and myocarditis.[34,35] *Chlamydophila* sp. infection is a cause of heart disease in reptiles. Granulomatous pericarditis and myocarditis was found in puff adders (*Bitis arietans*)[36] and necrotizing myocarditis in green turtles (*Chelonia mydas*).[37] Organisms could be demonstrated in macrophages in sections of paraffin-embedded heart. In emerald tree boas (*Corallus caninus*), there were histiocytic granulomas in the heart as well as in other organs. Small, basophilic organisms were seen on hematoxylin and eosin stained sections. Transmission electron microscopy of an intestinal granuloma demonstrated developmental stages of the chlamydial organisms.[38]

Mycoplasma infection of the heart was seen in American alligators (*Alligator mississippiensis*).[39] Rapidly growing mycoplasmas were identified by culture and polymerase chain reaction. The alligator isolate is a novel species in the mycoplasma family.

Spirorchid fluke (Digenea: Spirorchidae) infection was seen at necropsy of 96 stranded green turtles (*Chetonia mydas*).[40] Cardiovascular lesions included mural endocarditis, arteritis, and thrombosis, frequently accompanied by aneurysm formation. Spirorchiid trematode eggs were noted in the heart and other organs in black turtles (*Chelonia mydas agassizii*).[41] Seventy-five adult *Learedius learedi* were recovered from the heart of one of the turtles. During a 5-year period, 16 freshwater turtles (*Trachemys scripta elegans* and *Chrysemys picta*) died spontaneously.[42] Necropsy lesions included subcutaneous edema, hepatic necrosis, pancreatic necrosis, splenic necrosis, and intestinal parasites. Histologically, trematode eggs were seen within the myocardium, and in other tissues, associated with granulomatous lesions. The size and distribution of the eggs were consistent with *Spirorchis* sp. infection. Fluke infections can result in parasites within the three chambers of the heart and in major vessels (right aortic arch and brachiocephalic artery), attached to the walls, or free in the lumen.[43]

Whiptail lizards (*Cnemidophorus* spp.) from Texas and Colorado (USA), were examined for *Mesocestoides* sp. tetrathyridia and eleven (5%) were infected. Free tetrathyridia were found in the body cavity of lizards and encapsulated tetrathyridia were observed in the heart, liver, and stomach.[44]

Trachea and lungs

Infectious diseases have been considered the most commonly reported pulmonary diseases in reptiles in several surveys.[45-47]

Several viral diseases are associated with pneumonia. A paramyxovirus related to parainfluenza 2 (PI2) virus was recovered from the lungs of two dead Ottoman vipers (*Vipera xanthena xanthena*). Histologically there was interstitial pneumonia and degeneration and hyperplasia of bronchial and atrial epithelia. Scattered vacuoles, some of which contained eosinophilic inclusion bodies, were seen in the cytoplasm of several cells of affected epithelial tissues. Viral hemagglutination was inhibited by PI2 virus antiserum, but not by antisera to PI1, PI3, respiratory syncytial, or canine distemper viruses. Indirect immunofluorescence with PI2 viral antiserum specifically stained inclusions in the epithelial cells of respiratory tissues and infected cell cultures.[48]

In one study investigating respiratory disease in 80 boid snakes (family Boidae and Pythonidae), Ferlavirus (formerly known as ophidian paramyxovirus)-RNA was detected only in pythons. Inclusion body disease (IBD) was rarely seen in pythons but often in boas.[49] Eosinophilic intracytoplasmic inclusion bodies within the epithelial cells of the many tissues is diagnostic for inclusion body disease (IBD) of Boid snakes, an arenavirus.[50] This disease can be responsible for clinical signs including chronic regurgitation, incoordination, loss of righting reflexes, paresis, and an increased incidence of secondary infections such as stomatitis and pneumonia. Clinical signs referable to the central nervous system are more prominent in the Pythoninae subfamily.

Nidovirus infection resulted in pneumonia, tracheitis and esophagitis in ball pythons (*Python regius*). The histologic findings were of marked hyperplasia of epithelial cells lining air exchange areas (pneumocytes) with significant subacute interstitial inflammation and epithelial necrosis. Similar inflammatory and hyperplastic changes were also present in the trachea, esophagus and oral cavity.[51]

An adult male gopher tortoise (*Gopherus polyphemus*) had signs of upper respiratory disease. On necropsy a severe, extensive necrotizing ulcerative tracheitis, multifocal necrotizing pneumonia, and multifocal necrotizing ulcerative pharyngitis and esophagitis were seen. Intracytoplasmic basophilic inclusions occurred within necrotic epithelial cells. On transmission electron microscopy virions and cytoplasmic inclusions were morphologically similar to those of the Family Iridoviridae.[52] In other animals with iridovirus (Russian tortoise [*Testudo horsfieldii*], and a box turtle [*Terrapene carolina*]) no lesions were described; however, a captive Burmese star tortoise (*Geochelone platynota*), a wild gopher tortoise, Eastern mud turtle (*Kinosternon subrubrum*) and 5 Eastern box turtles (*Terrapene carolina carolina*) were found to be infected with ranavirus. Several animals had varying degrees of multicentric vasculitis or thrombosis, necrosis of hematopoietic tissues, as well as multifocal necrotizing tracheitis. In some cases, basophilic intracytoplasmic inclusion bodies were observed within epithelial cells of the oral mucosa, esophagus, stomach, and trachea.[53,54]

Herpesvirus-associated rhinitis, stomatitis, glossitis, tracheitis and bronchopneumonia has been described in a number of tortoises (California desert tortoise [*Gopherus agassizii*],[55] spur-thigh tortoises [*Testudo graeca*],[56] Argentine tortoises [*Chelonoidis chilensis*], Mediterranean land tortoises [*Testudo* species], Hermann's [*Testudo hermanni*] and four-toed tortoises [*Testudo horsfieldii*]). This herpesviral infection is frequently complicated by secondary bacterial infections and generally results in severe debilitation of the tortoise. The transmission is most likely from infected carriers although the pathogenesis has not been fully described.

Herpesvirus infection was also diagnosed in 14 juvenile (15- to 20-month-old) green turtles (*Chelonia mydas*) with clinical signs of respiratory tract disease. Gross lesions included periglottal necrosis, tracheitis with intraluminal caseous and laminated necrotic debris, and severe pneumonia. Microscopically, the turtles had fibrinonecrotic inflammation around the glottal opening, tracheitis, and severe bronchopneumonia and interstitial pneumonia. Multifocally, tracheal epithelial cells adjacent to areas of necrosis were karyomegalic and had amphophilic intranuclear inclusions. A secondary bacterial infection was noted. Ultrastructurally, intranuclear viral particles (88-99 nm in diameter) were seen. The particles most closely resembled those of herpesviruses.[57]

Atlantic loggerhead sea turtles (*Caretta caretta*) recovered by a rescue group had gross and histologic evidence of a herpesviral infection consisting of oral, respiratory, cutaneous, and genital lesions characterized by necrosis, ulceration, syncytial cell formation, and intranuclear inclusion bodies. By nested polymerase chain reaction two unique herpesviral sequences referred to as loggerhead genital-respiratory herpesvirus and loggerhead orocutaneous herpesvirus were identified.[58]

Paramyxoviruses (PMVs) respiratory disease have only been documented in tortoises in a very few cases. A leopard tortoise (*Geochelone pardalis babcocki*) presented with lethargy, respiratory disease, and large amounts of mucopurulent discharge from the nares and oral cavity. On gross pathology there was severe diffuse consolidation of the lungs which were pale, firm, and filled with a thick serous exudate. By molecular methods (polymerase chain reaction [PCR]) several tissues were positive for Paramyxoviruses. Sequence analysis of the PMV amplicons revealed that the tortoise was coinfected with at least two different squamatid PMV species and not with a chelonian host-specific PMV.[59]

Infection with West Nile virus (WNV) in addition to myocardial degeneration with necrosis produced a mild interstitial pneumonia and lesions in many other organs in alligators.[31] Immunohistochemistry identified WNV antigen in the tissues.

In one review gram-negative bacterial infections were considered the most common cause of pneumonia in reptiles.[45] Bacterial pneumonia can lead to systemic complications. Poor husbandry conditions are an important trigger for the development of respiratory signs and pneumonia in snakes. Different bacterial pathogens are commonly isolated in almost all snakes with pneumonia, with Salmonella species being the most common in boid snakes.[49] In a group of sea kraits (*Laticauda* spp.), nine died of sepsis secondary to necrotizing enteritis or pneumonia. Based on the clinical picture, it was considered that stress, such as transport, captivity, or possible concurrent viral infection, resulted in a septic event and death.[61]

Bacterial pneumonia was one of the most frequently encountered clinically significant lesions in Kemp's ridley sea turtles (*Lepidochelys kempii*).[62]

In a group of California desert tortoises (*Gopherus agassizii*), 5 of 7 with respiratory disease had mycoplasmosis based on the presence of chronic proliferative rhinitis and positive serologic tests and/or isolation of *Mycoplasma* species.[63] Mycoplasma has also caused proliferative lymphocytic tracheitis and pneumonia in a Burmese python (*Python molurus bivittatus*). Polymerase chain reaction analysis of the 16S rRNA gene sequence indicated 0.90 similarity to *Mycoplasma agassizii*, an organism previously shown to cause respiratory disease in reptiles.[64] Pneumonia has been seen as a part of systemic disease due to *Chlamydophilia* sp. infection in puff adders (*Bitis arietans*).[36]

Mycobacterial infection can also cause pneumonia in reptiles.[23,45,65,66] It is considered a sporadic disease in captive collections, usually seen as a chronic disease. A variety of mycobacterial genera have been isolated, most of which are generally environmental microbes.[65] These infections are commonly disseminated with the portal of entry suspected to be oral or through defects in the skin (trauma). The lesions are typical for mycobacterium with granulomatous inflammation.

Mycotic pneumonia has been considered an occasional but uncommon diagnosis in reptiles,[23] with a variety of organisms isolated. Suboptimal temperatures[21] can lead to mycotic pneumonia in captive chelonians. Phaeohyphomycosis (pigmented fungi) resulted in obstructive tracheitis in stranded green sea turtles, *Chelonia mydas*.[67]

Other reported causes include mycotic pneumonia associated with *Aspergillus* sp. in green anacondas (*Eunectes murinus*),[68] and chronic pneumonia in a desert tortoise with respiratory disease.[63] In snakes, most cases of systemic mycosis begin in the lungs and then disseminate.[69]

Protozoal infections of the lung can occur in reptiles. Microsporidia, several cases having *Encephalitozoon*-like morphology, have been reported in bearded dragons (*Pogona vitticeps*).[70,71] These lizards presented with nonspecific signs of illness. On histology clusters of light basophilic intracytoplasmic microorganisms are present and free microbes in areas of necrosis. The microorganisms were within cytoplasmic vacuoles in hepatic, renal, adrenal gland, splenic, pulmonary, enteric and gastric epithelial cells, capillary endothelial cells, and ventricular ependymal cells in the brain. The microbes were associated with granulomatous inflammation. The microorganism is Gram's stain positive, acid fast stain positive, and has a small polar granule that stained using the periodic acid-Schiff reaction. Electron microscopy revealed merogonic and sporogonic stages of a protozoa compatible with members of the phylum Microspora.[71] Nucleotide sequencing revealed high similarity with published *E. cuniculi* sequences in two cases.

An intranuclear coccidia has been seen in the lung of several species of captive tortoises in the United States, leading to severe proliferative pneumonia in some cases.[21]

Several types of reptile metazoan parasites have at least a portion of their life-cycle in the lung.[23] Nematodes of the genus *Rhabdia* lay eggs within the pulmonary parenchyma. If the eggs rupture, there can be a marked inflammatory response.

Lungworm infection was seen in loggerhead sea turtles (*Caretta caretta*). The lungworms may not have been the primary cause of illness, and they may be only contributory or incidental. Lesions in those that died included tracheal and bronchial epithelial hyperplasia and goblet cell hyperplasia. Lesions caused directly by the parasites seem to be restricted to the upper respiratory tract; debris produced by the worms may lead to changes in the lungs.[72]

Pentastomids (phylum Pentastomida) comprise a highly specialized taxon of arthropod-like parasites that are suspected to share a common ancestry with branchiuran crustaceans. Pentastomids are long-lived, often large parasites that probably became adapted to the lungs of amphibians and reptiles early in their evolutionary history. The majority of adult pentastomids feed on blood in the lungs and trachea of the definitive reptilian hosts and cause minimal damage. After mating once, the females deposit a large number of embryonated eggs. The eggs are carried up the trachea by the ciliary action of the mucosal lining to the oral cavity. The eggs are then swallowed and pass in the feces. In the typical life cycle, an invertebrate (e.g., cockroach) or vertebrate intermediate host (e.g., rodents, herbivores, carnivores, and primates) ingests the eggs. The larvae hatch and encyst within tissues where they will molt several times to an infective nymph. When the definitive reptile host eats the intermediate host, the nymph completes the life cycle by developing into an adult in the lung. In aberrant hosts and in the normal hosts under stress, the parasites can be associated with inflammation. If there is significant inflammation, air flow obstruction as well as secondary bacterial infections can develop.

Two Bosc's monitor lizards (*Varanus exanthematicus*) developed clinical signs of pentastomiasis. One died of chronic parasitic pneumonia associated with adult pentastomids of an undescribed *Sambonia* sp. Eggs and immature pentastomids were also seen in histologic sections of the lungs and liver. The other animal had fragments of pentastomid larvae in a laryngeal biopsy, and the animal recovered after treatment with ivermectin and supportive therapy.[73]

Several genera of trematodes can cause lesions in the lower respiratory tract of reptiles.[23] Identification of eggs from lung or tracheal washes is necessary for an exact etiologic diagnosis.

Esophagus and stomach

Adenovirus infections are documented in at least twelve different species of reptiles. In contrast to their mammalian and avian counterparts, reptilian adenoviruses are not well characterized as to their pathogenic potential and their ability to cause primary disease.

Adenoviral infections of bearded dragons are common and this viral infection has become established in certain US lizard breeding groups of the genus *Pogona* (bearded dragons).[74-76] The common clinical signs in young dragons are a progressive weakness, anorexia, circling, and head tilt.[74] The lesions are of severe acute coagulative hepatocellular necrosis and the large intranuclear basophilic inclusion bodies in the hepatocytes and enterocytes. It is speculated that there is possible vertical transmission through the egg in utero or at time of oviposition. The adenovirus infections are also reported to be associated with a second virus, Dependovirus.[75] This virus is defective and appears to need the adenovirus to replicate. It is not associated with a disease condition.

Adenovirus was associated with proliferation of tracheal and esophageal mucosa in a 6-month-old Jackson's chameleon (*Chamaeleo jacksoni*). The clinical signs included opisthotonus and anorexia. Histologically, eosinophilic intranuclear inclusion bodies found within ciliated epithelial cells of the esophageal and tracheal mucosa. The inclusions were morphologically consistent with adenovirus by electron microscopy.[77]

A group of seven juvenile California mountain kingsnakes (*Lampropeltis zonata multicincta*) developed severe acute gastroenteritis. The clinical signs included emesis, dehydration, and sudden death. Histopathology demonstrated segmental acute mucosal necrosis and hyperplasia. The enterocytes had intranuclear inclusions consistent with adenovirus.[78]

Esophageal adenovirus-like infection was described in a palm viper (*Bothriechis marchi*) with a concurrent inclusion body-like disease using electron microscopy. The adult snake was captive-born and found dead. The lesions included a stomatitis and esophagitis.[79]

Inclusion body disease (IBD) of boid snakes can be histologically recognized in lining mucosal epithelial and lymphoid cells aggregates of the esophagus, as well as within the mucosal cells of the stomach and less frequently in the intestines.[80] The eosinophilic intracytoplasmic inclusions have been described in boas and pythons (family Boidae), palm vipers (*Bothriechis marchi*) and corn snakes (*Elaphe guttata*).[81-83] It is important to note that these inclusions do not always confirm IBD, as the inclusions in the corn snakes did not have consistent morphology when examined by electron microscopic.[82] This disease can be responsible for clinical signs including chronic regurgitation, incoordination, loss of righting reflexes, paresis, and an increased incidence of secondary infections such as stomatitis and pneumonia. Clinical signs referable to the central nervous system are more prominent in the Pythoninae subfamily.[84,85] Antemortem diagnostics include CBC's and biopsies of the esophagus (especially of lymphocytic stromal aggregates or tonsils), gastric mucosa, and liver.[64] The etiologic agent of IBD is an arenavirus.[50]

Cryptosporidiosis is a well-known gastrointestinal disease of snakes and lizards. Hypertrophic or proliferative gastritis is a common manifestation of *Cryptosporidium* infections in snakes, and atropic gastritis has been described in lacertas. Polymerase chain reaction (PCR) has detected *Cryptosporidium varanii* (*saurophilum*) in corn snakes (*Pantherophis guttatus*) and in leopard geckos (*Eublepharis macularius*). *Cryptosporidium serpentis* was found in other leopard geckos. A *Cryptosporidium* sp. lizard genotype was reported in one leopard gecko and one corn snake.[86] Pseudoparasitic cryptosporidian species from those ingested with the prey (pseudoparasites) have been described in snakes.[86]

Two Russian tortoises (*Agrionemys* [*Testudo*] *horsfieldii*) and a pancake tortoise (*Malacochersus tornieri*), all from separate collections had histologic evidence of intestinal cryptosporidiosis and one with gastric cryptosporidiosis. A mild inflammation was associated with the infections; however, only one case of the intestinal case had mucosal hyperplasia. Consensus *Cryptosporidium* sp. PCR and sequencing was used to characterize the *Cryptosporidium* sp. present in these three tortoises. This may be an under-recognized problem in tortoises.[87]

The coccidia *Isopora amphiboluri* is most common in bearded dragons (*Pogona vitticeps)* and can be associated with clinical signs.[88] It has a direct life cycle and can predispose to other more serious diseases. An adverse synergistic effect of the coccidiosis with the adenoviral infection may be responsible for increased morbidity and mortality.[88]

Penned green turtles (*Chelonia mydas*) became infected with a larval nematode (*Anisakis sp.* Type I) most likely from the food source, sardine (*Harengula ovalis*).[89] The larvae were associated with hemorrhagic ulcers in the pyloroduodenal junction of the alimentary tract.

Neoplasia

Thyroid adenomatous hyperplasia and follicular adenomas have been reported in lizard, snake, and turtle species,[90-93] whereas thyroid carcinomas in reptiles are rare. Reports of thyroid carcinomas have been described in a Chinese crocodile lizard (*Shinisaurus crocodilurus*),[94] a red-eared slider (*Trachemys scripta elegans*),[95] an Indian black turtle (*Melanochelys trijuga*), and a rough and smooth knob-tail geckos (*Nephrurus amyae* and *N. levis*).[90] The clinical signs in the lizards included an intraoral mass, ventral neck swelling, oral hemorrhage, and weight loss. The thyroid carcinoma metastasized to the liver and lungs in the rough knob-tail gecko (*Nephrurus amyae*).[90] The functional aspects of the tumor are rarely described either by blood hormone levels or clinical signs.[90] One iguana (*Iguana iguana*) with a thyroid adenoma presented with polyphagia, aggression, and tachycardia. The elevated total T4 (30.0 nmol/L) resolved to a euthyroid state after surgical removal.[91]

A C cell carcinoma of the thyroid gland in a crocodile lizard (*Shinisauris crocodilurus*) metastasized to the liver and mesentery.[96]

Primary tumors of the reptile cardiovascular system are not common. Among those reported are a rhabdomyosarcoma, in a boa constrictor (*Boa constrictor*), and a fibrosarcoma in a Gaboon viper (*Bitis gabonica*).[97] Other tumors that may be primary or be part of a disseminated process include lymphoblastic malignant lymphoma and lymphoid leukemias in several species of reptiles.[97,98] Thirty-nine percent of green turtles (*Chelonia mydas*) with fibropapillomatosis (FP) had disseminated internal tumors, most of them in the lung, kidney, and heart.[99] Fibrosarcomas of low-grade malignancy were most frequently noted in the heart, and heart tumors had a predilection for the right atrium. A disseminated mast cell tumor diagnosed in an eastern kingsnake (*Lampropeltis getulus getulus*) involved the heart and numerous other organs.[100] A disseminated chondrosarcoma involved the heart in a corn snake (*Elaphe guttata*).[101] Large areas of myocardium had been replaced by neoplastic tissue. A metastatic oviductal adenocarcinoma has also been reported in the heart of a corn snake.[102]

In one survey[97] two primary adenocarcinomas of the respiratory tract were listed, one of tracheal origin and the other a bronchogenic carcinoma. Tracheal chondromas seem to have a predilection for ball pythons (*Python regius*).[103-105] Adult ball pythons generally present for severe dyspnea. Partial obstructions of the tracheal lumen can be identified radiographically and/or visualized. Histologically, the lesions are tracheal ring chondromas and appear to be benign. Surgical removal and tracheal resection was successful in several cases.

Disseminated tumors such as lymphoma can involve the lung,[97] and a variety of metastatic tumors have been reported. These include fibromas,[99] oviductal carcinoma,[102] chondrosarcoma,[101] squamous cell carcinoma,[106] fibrosarcoma,[107] and a plasma cell tumor.[108]

The majority of the adenocarcinoma/carcinoma gastrointestinal tract tumors are reported in snakes and rarely in lizards and chelonians.[109-112] These tumors will result in variable clinical signs of the digestive tract including constipation, regurgitation, and progressive enlargement of the coelom. In snakes the differential should include gastric cryptosporidia infections.

There are several reports of a highly malignant gastric neuroendocrine carcinomas in young bearded dragons (*Pogona vitticeps*).[113,114] These tumors readily metastasize to the liver. The common clinical signs are anorexia, vomiting, hyperglycemia, and anemia. The author has identified metastases from gastric carcinomas into the intestine and liver of a carpet python (*Morelia spilota*) and in the liver of two bearded dragons (*Pogona vitticeps*).

References

1. Mitchell MA. Reptile cardiology. *Vet Clin North Am Exot Anim Pract.* 2009;12:65-79.
2. Rush EM, Donnelly TM, Walberg J. Aneurysm and subsequent cardiopulmonary arrest in a Burmese python (*Python molurus bivittatus*). *Proc Annu Conf Am Assoc Zoo Vet.* 1999;134-138.
3. Kik MJL, Mitchell MA. Reptile cardiology: a review of anatomy and physiology, diagnostic approaches, and clinical disease. *Semin Avian Exot Pet Med.* 2005;14:52-60.
4. Pizzi R, Martinez Pereira Y, Feltner Rambaud Y, et al. Secundum atrial septal defect in a Komodo dragon (*Varanus komodoensis*). *Vet Rec.* 2009;164:472-473.
5. Jensen B, Wang TW. Hemodynamic consequences of cardiac malformations in two juvenile ball pythons (*Python regius*). *J Zoo Wildl Med.* 2009;40:752–756.
6. Brockman HL, Kennedy JP. Interventricular septal defect in *Alligator mississipiensis* Daudin. *Tex Rep Biol Med.* 1962;20:719-720.
7. Buffetaut E, Li J, Tong H, Zhang H. A two-headed reptile from the Cretaceous of China. *Biol Lett.* 2007;3(1):80-81.
8. Andreadis PT, Burghardt GM. Feeding behavior and an oropharyngeal component of satiety in a two-headed snake. *Physiol Behav.* 1993;54(4):649-658.
9. Palmieri C, Selleri P, Di Girolamo N, Montani A, Salda LD. Multiple congenital malformations in a dicephalic spur-thighed tortoise (*Testudo graeca ibera*). *J Comp Pathol.* 2013;149(2-3):368-371.
10. Orós J, Rodriguez JL, de Los Monteros AE, Rodríguez F, Herráez P, Fernández A. Tracheal malformation in a bicephalic Honduran milk snake (*Lampropeltis hondurensis*) and subsequent fatal Salmonella arizonae infection. *J Zoo Wildl Med.* 1997;28(3):331-335
11. Jacobson ER, Berry KH, Stacy B, Huzella LM, Kalasinsky VF, Fleetwood ML, Mense MG. Oxalosis in wild desert tortoises, *Gopherus agassizii*. *J Wildl Dis.* 2009;45(4):982-988.
12. Fleming GJ, Heard DJ, Uhl EW, Johnson CM. Thymic hyperplasia in subadult Galapagos tortoises, *Geochelone nigra*. *J Herpetol Med Surg.* 2004;14(1):24-27.
13. Rush EM, Donnelly TM, Walberg J. Aneurysm and subsequent cardiopulmonary arrest in a Burmese python (*Python molurus bivittatus*). *Proc Annu Conf Am Assoc Zoo Vet.* 1999;134-138.

14. Schuchman SM, Taylor DO. Arteriosclerosis in an iguana (*Iguana iguana*). *J Am Vet Med Assoc*. 1970;157:614-616.

15. Barten S. Cardiomyopathy in a king snake (*Lampropeltis calligaster rhombomaculata*). *Vet Med Small Anim Clin*. 1980;75:125-129.

16. Jacobson ER, Seely JC, Novilla MN, et al. Heart failure associated with unusual hepatic inclusions in a Deckert's rat snake. *J Wildl Dis*. 1980;15:75-81.

17. Clippinger TL. 1993. Aortic stenosis and atrio/ventricular dilatation in a green iguana (*Iguana iguana*). *Proc Annu Conf Am Assoc Zoo Vet*. 1993;342-344.

18. Frye FL. Common pathologic lesions & disease processes. In: Frye FL, ed. *Biomedical and Surgical Aspects of Captive Reptile Husbandry*. Malabar, FL: Krieger Publishing;1991:560.

19. Jacobson ER. Gentamicin related visceral gout in two boid snakes. *Vet Med Small Anim Clin*. 1976;71: 361-363.

20. Montali RJ, Bush M, Smeller JM. The pathology of nephrotoxicity of gentamicin in snakes. *Vet Pathol*. 1979;16:108-111.

21. Origgi FC, Jacobson ER. Diseases of the respiratory tract of chelonians. *Vet Clin North Am Exot Anim Pract*. 2000;3:537-549.

22. Schmidt RE, Hubbard GB. Respiratory system. In: Schmidt RE, Hubbard GB, eds. *Atlas of Zoo Animal Pathology*. Boca Raton, FL: CRC Press;1987:19.

23. Murray MJ. Pneumonia and lower respiratory tract disease. In: Mader DR, ed. *Reptile Medicine and Surgery*. Philadelphia, PA: WB Saunders;2006:865-877.

24. Benson KG. Reptilian Gastrointestinal diseases. *Semin Avian Exotic Pet Med*. 1999;8(2):90-97.

25. Mitchell MA, Diaz-Figueroa O. Clinical reptile gastroenterology. *Vet Clin North Am Exot Anim Pract*. 2005;8(2):277-298.

26. Adams CJ, Sleeman JM. Clinical challenge. The coelomic cavity consistent with a foreign body. *J Zoo Wildl Med*. 2005;36(4):731-734.

27. Klaphake E, Cross C, Patton S, Head L. Gastric impaction in a milk snake, *Lampropeltis triangulum*, caused by *Kalicephalus* sp. *J Herpetol Med Surg*. 2005;15(1):21-23.

28. Hasbun CR, Lawrence AJ, Samour JH, al-Ghais S. Duodenal volvulus in free-living green turtles from coastal United Arab Emirates. *J Wildl Dis*. 1998;34(4):797-800.

29. Rogers EP, Windsor RS. Gastric impaction in captive crocodiles (*Crocodilus niloticus*). *J S Afr Vet Assoc*. 1982;53(4):254.

30. Murray MJ. Cardiology. In: Mader DR, ed. *Reptile Medicine and Surgery*. Philadelphia, PA: WB Saunders;2006:181-195.

31. Jacobson ER, Ginn PE, Troutman JM, et al. West Nile virus infection in farmed American alligators (*Alligator mississippiensis*) in Florida. *J Wildl Dis*. 2005;41:96-106.

32. Obendorf DL, McManus CJ. *Vibrio damsela* infection in a stranded leatherback turtle (*Dermochelys coriacea*). *J Wildl Dis*. 1987;23:666-668.

33. Jacobson ER, Homer B, Adams W. Endocarditis and congestive heart failure in a Burmese python (*Python molurus bivittatus*). *J Zoo Wildl Med*. 1991;22:245-248.

34. Innis CJ. Myocardial abscess and hemopericardium in a green iguana (*Iguana iguana*). *Proc Annu Conf Assoc Rept Amph Vet*. 2000;185-188.

35. Gray TZ. Cranial body swelling in a boa constrictor, *Boa constrictor constrictor*. *J Herpetol Med Surg*. 2004;14(4):22-24.

36. Jacobson ER, Gaskin JM, Mansell J. Chlamydial infection in puff adders (*Bitis arietans*). *J Zoo Wildl Med*. 1989;20:364-369.

37. Homer BL, Jacobson ER, Schumacher J, Scherba G. Chlamydiosis in mariculture-reared green sea turtles (*Chelonia mydas*). *Vet Pathol*. 1994;31:1-7.

38. Jacobson E, Origgi F, Heard D, Detrisac C. Immunohistochemical staining of chlamydial antigen in emerald tree boas (*Corallus caninus*). *J Vet Diagn Invest*. 2002;14: 487-494.

39. Clippinger TL, Bennett RA, Johnson CM, et al. Morbidity and mortality associated with a new mycoplasma species from captive American alligators (*Alligator mississippiensis*). *J Zoo Wildl Med*. 2000;31:303-314.

40. Gordon N, Kelly WR, Cribb TH. Lesions caused by cardiovascular flukes (Dogemea:spirorchidae) in stranded green turtles (*Chetonia mydas*). *Vet Pathol*. 1998;35:21-30.

41. Cordero-Tapia A, Gardner SC, Arellano-Blanco J, Inohuye-Rivera RB. Learedius learedi infection in black turtles (*Chelonia mydas agassizii*), Baja California Sur, Mexico. *J Parasitol*. 2004;90:645-647.

42. Johnson CA, Griffith JW, Tenorio P, et al. Fatal trematodiasis in research turtles. *Lab Anim Sci*. 1998;48: 340-343.

43. Glazebrook JS, Campbell RS, Blair D. Studies on cardiovascular fluke (Digenea: Spirorchiidae) infections in sea turtles from the Great Barrier Reef, Queensland, Australia. *J Comp Pathol*. 1989;101:231-250.

44. McAllister CT, Cordes JE, Conn DB, Singleton J, Walker JM. Helminth parasites of unisexual and bisexual whiptail lizards (Teiidae) in North America. V. *Mesocestoides* sp. tetrathyridia (Cestoidea: Cyclophyllidea) from four species of Cnemidophorus. *J Wildl Dis*. 1991;27:494-497.

45. Barker IK, Goltz JP. Pneumonia in reptiles: Clinico-pathologic correlations. *Proc Annu Conf Am Assoc Zoo Vet*. 1980;60-63.

46. Keymer IF. Diseases of chelonians: (1) necropsy survey of tortoises. *Vet Rec*. 1978;103:548-552.

47. Keymer IF. Diseases of chelonians: (2) necropsy survey of terrapins and turtles. *Vet Rec*. 1978;103:577-582.

48. Potgieter LN, Sigler RE, Russell RG. Pneumonia in Ottoman vipers (*Vipera xanthena xanthena*) associated with a parainfluenza 2-like virus. *J Wildl Dis*. 1987;23:355-360.

49. Schmidt V, Marschang RE, Abbas MD, Ball I, Szabó I, Helmuth R, Plenz B, Spergser J, Pees M. Detection of pathogens in Boidae and Pythonidae with and without respiratory disease. *Vet Rec*. 2013;172(9):236.

50. Stenglein MD, Jacobson ER, Chang L-W, Sanders C, Hawkins MG, Guzman DS-M, Reavill DR, et al. Widespread recombination, reassortment, and transmission of unbalanced compound viral genotypes in natural arenavirus infections. *PLoS Pathog*. 2015;11(5): e1004900.

51. Uccellini L, Ossiboff RJ, de Matos REC, Morrisey JK, Petrosov A, Navarrete-Macias I, Jain K, Hicks AL, Buckles EL, Tokarz R, McAloose D, Lipkin WI. Identification of a novel nidovirus in an outbreak of fatal respiratory disease in ball pythons (*Python regius*). *Virol J*. 2014;11:144

52. Westhouse RA, Jacobson ER, Harris RK, et al. Respiratory and pharyngo-esophageal iridovirus infection in a gopher tortoise (*Gopherus polyphemus*). *J Wildl Dis*. 1996;32:682-686.

53. Winzeler ME, Hamilton MT, Tuberville TD, Lance SL. First case of ranavirus and associated morbidity and mortality in an eastern mud turtle *Kinosternon subrubrum* in South Carolina. *Dis Aquat Organ*. 2015;114(1):77-81.

54. Johnson AJ, Wellehan JFX, Pessier AP. Iridovirus infections of turtles and tortoises. *Proc Annu Conf Am Assoc Zoo Vet*. 2004;11-12.

55. Pettan-Brewer KC, Drew ML, Ramsay E, et al. Herpesvirus particles associated with oral and respiratory lesions in a California desert tortoise (*Gopherus agassizii*). *J Wildl Dis*. 1996;32:521-526.

56. Muro J, Ramis A, Pastor J, Velarde R, Tarrés J, Lavin S. Chronic rhinitis associated with herpesviral infection in captive spur-thighed tortoises from Spain. *J Wildl Dis*. 1998;34(3):487-495.

57. Jacobson ER, Gaskin JM, Roelke M. Conjunctivitis, tracheitis, and pneumonia associated with herpesvirus infection in green sea turtles. *J Am Vet Med Assoc*. 1986;189:1020-1023.

58. Stacy BA, Wellehan JFX, Foley AM, Coberley SS, Herbst LH, Manire CA, Garner MM, Brookins MD, Childress AL, Jacobson ER. Two herpesviruses associated with disease in wild Atlantic loggerhead sea turtles (*Caretta caretta*). *Vet Microbiol*. 2008;126(1-3):63-73.

59. Papp T, Seybold J, Marschang RE. Paramyxovirus infection in a leopard tortoise (*Geochelone pardalis babcocki*) with respiratory disease. *J Herpetol Med Surg*. 2010;20(2-3):64-68.

60. Jacobson ER, Ginn PE, Troutman JM, et al. West Nile virus infection in farmed American alligators (*Alligator mississippiensis*) in Florida. *J Wildl Dis*. 2005;41:96-106.

61. Chinnadurai SK, Brown DL, Van Wettere A, et al. Mortalities associated with sepsis, parasitism, and disseminated round cell neoplasia in yellow-lipped sea kraits (*Laticauda colubrina*). *J Zoo Wildl Med*. 2008;39:626-630.

62. Innis C, Nyaoke AC, Williams CR 3rd, et al. Pathologic and parasitologic findings of cold-stunned Kemp's ridley sea turtles (*Lepidochelys kempii*) stranded on Cape Cod, Massachusetts, 2001-2006. *J Wildl Dis*. 2009;45:594-610.

63. Homer BL, Berry KH, Brown MB, et al. Pathology of diseases in wild desert tortoises from California. *J Wildl Dis*. 1998;34:508-523.

64. Penner JD, Jacobson ER, Brown DR. A novel *Mycoplasma* sp. associated with proliferative tracheitis and pneumonia in a Burmese python (*Python molurus bivittatus*). *J Comp Pathol*. 1997;117:283-288.

65. Jacobson ER. Bacterial diseases of reptiles. In: Jacobson ER, ed. *Infectious Diseases and Pathology of Reptiles*. Boca Raton, FL: CRC Press;2007:461-526.

66. Woodhouse SJ, Fitzgerald SD, Lim A, Bolin SR. Disseminated *Mycobacterium haemophilum* infection in an ASSAM trinket snake (*Elaphe frenata*). *J Zoo Wildl Med*. 2014;45(4):966-969.

67. Origgi FC, Jacobson ER. Diseases of the respiratory tract of chelonians. *Vet Clin North Am Exot Anim Pract*. 2000;3:537-549.

67. Donnelly K, Waltzek TB, Wellehan JFX, Jr., Sutton DA, Wiederhold NP, Stacy BA. Phaeohyphomycosis resulting in obstructive tracheitis in three green sea turtles *Chelonia mydas* stranded along the Florida coast. *Dis Aquat Organ*. 2015;113(3):257-262.

68. Miller DL, Radi ZA, Stiver SL, et al. Cutaneous and pulmonary mycosis in green anacondas (*Euncectes murinus*). *J Zoo Wildl Med*. 2004;35:557-561.

69. Jacobson ER. Mycotic diseases of reptiles. In: Jacobson ER, ed. *Infectious Diseases and Pathology of Reptiles*. Boca Raton, FL: CRC Press;2007:527-570.

70. Richter B, Csokai J, Graner I, Eisenberg T, Pantchev N, Eskens HU, Nedorost N. Encephalitozoonosis in two inland bearded dragons (*Pogona vitticeps*). *J Comp Pathol*. 2007;148(2-3):278-282.

71. Jacobson ER, Green DE, Undeen AH, et al. Systemic microsporidiosis in inland bearded dragons (*Pogona vitticeps*). *J Zoo Wildl Med*. 1998;29:315-323.

72. Manire CA, Kinsel Jm, Anderson ET, et al. Lungworm infection in three loggerhead sea turtles, *Caretta caretta*. *J Zoo Wildl Med*. 2008;39:92-98.

73. Flach EJ, Riley J, Mutlow AG, et al. Pentastomiasis in Bosc's monitor lizards (*Varanus exanthematicus*) caused by an undescribed Sambonia species. *J Zoo Wildl Med*. 2000;31:91-95.

74. Kim DY, Mitchell MM, Bauer RW, Poston R, Cho D-Y. An outbreak of adenoviral infection in inland bearded dragons (*Pogona vitticeps*) coinfected with dependovirus and coccidial protozoa (*Isospora* sp.). *J Vet Diagn Invest*. 2002;14:332-334.

75. Jacobson ER, Kopitf W, Kennedy A, Funk RS. Coinfection of a bearded dragon, *Pogona vitticeps*, with Adenovirus- and Dependovirus-like viruses. *Vet Pathol*. 1996;33:343-346.

76. Moormann S, Seehusen F, Reckling D, Kilwinski J, Puff C, Elhensheri M, Wohlsein P, Peters M. Systemic Adenovirus infection in bearded dragons (*Pogona vitticeps*): histological, ultrastructural and molecular findings. *J Comp Pathol.* 2009;141:78-83.

77. Jacobson ER, Gardiner CH. Adeno-like virus in esophageal and tracheal mucosa of a Jackson's chameleon (*Chamaeleo jacksoni*). *Vet Pathol.* 1990;27(3):210-212.

78. Wozniak EJ, DeNardo DF, Brewer A, Wong V, Tarara RP. Identification of Adenovirus- and Dependovirus-like agents in an outbreak of fatal gastroenteritis in captive born California mountain kingsnakes, *Lampropeltis zonata multicincta. J Herpetol Med Surg.* 2000;10(2):4-7.

79. Raymond JT, Garner MM, Murray S, Nordhausen R. Oroesophageal Adenovirus-like infection in a palm viper, *Bothriechis marchi*, with inclusion body-like disease. *J Herpetol Med Surg.* 2002;12(3):30-32.

80. Garner MM, Raymond JT. Methods for diagnosing inclusion body disease in snakes. *Exotic DVM.* 2004;6(3):90-92.

81. Vancraeynest D, Pasmans F, Martel A, Chiers K, Meulemans G, Mast J, Zwart P, Ducatelle R. Inclusion body disease in snakes: a review and description of three cases in boa constrictors in Belgium. *Vet Rec.* 2006;158(2):757-760.

82. Fleming GJ, Heard DJ, Jacobson ER, Buergelt C. Cytoplasmic inclusions in corn snakes, *Elaphe guttata*, resembling inclusion body disease of boid snakes. *J Herpetol Med Surg.* 2003;13(2):18-22.

83. Raymond JT, Garner MM, Nordhausen W, Jacobson ER. A disease resembling inclusion body disease of boid snakes in captive palm vipers (*Bothriechis marchi*). *J Vet Diagn Invest.* 2001;13:82-86.

84. Schumacher J, Jacobson ER, Homer BL, Gaskin JM . Inclusion body disease in boid snakes. *J Zoo Wildl Med.* 1994;25:511-524.

85. Carlisle-Nowak MS, Sullivan N, Carrigan M, Knight C, Ryan C, Jacobson ER. Inclusion body disease in two captive Australian pythons (*Morelia spilota variegata* and *Morelia spilota spilota*). *Aust Vet J.* 1998;76(2):98-100.

86. Richter B, Nedorost N, Maderner A, Weissenbock H. Detection of Cryptosporidium species in feces or gastric contents from snakes and lizards as determined by polymerase chain reaction analysis and partial sequencing of the 18S ribosomal RNA gene. *J Vet Diag Invest.* 2011;23(3):430-435.

87. Griffin C, Reavill DR, Stacy BA, Childress AL, Wellehan Jr. JFX. Cryptosporidiosis caused by two distinct species in Russian tortoises and a pancake tortoise. *Vet Parasitol.* 2010;170:14-19.

88. Kim DY, Mitchell MM, Bauer RW, Poston R, Cho D-Y. An outbreak of adenoviral infection in inland bearded dragons (*Pogona vitticeps*) coinfected with Dependovirus and coccidial protozoa (*Isospora* sp.). *J Vet Diagn Invest.* 2002;14:332-334.

89. Burke JB, Rodgers LJ. Gastric ulceration associated with larval nematodes (*Anisakis* sp. type I) in pen reared green turtles (*Chelonia mydas*) from Torres Strait. *J Wildl Dis.* 1982;18:41-46.

90. Hadfield CA, Clayton LA, Clancy MM, Beck SE, Mangus LM, Montali RJ. Proliferative thyroid lesions in three diplodactylid geckos: *Nephrurus amyae*, *Nephrurus levis*, and *Oedura marmorata*. *J Zoo Wildl Med.* 2012;43(1):131-140.

91. Hernandez-Divers SJ, Knott CD, MacDonald J. Diagnosis and surgical treatment of thyroid adenoma-induced hyperthyroidism in a green iguana (*Iguana iguana*). *J Zoo Wildl Med.* 2001;32(4):465-475.

92. Garner MM, Hernandez-Divers SM, Raymond JT. Reptile neoplasia: a retrospective study of case submissions to a specialty diagnostic service. *Vet Clin North Am Exot Anim Pract.* 2004;7(3):653-671.

93. Frye FL. Common pathologic lesions and disease processes. In: *Biomedical and Surgical Aspects of Captive Reptile Husbandry*. Malabar, FL: Krieger Publishing;1991:529-619.

94. Whiteside DP, Garner MM. Thyroid adenocarcinoma in a crocodile lizard, *Shinisaurus crocodilurus*. *J Herpetol Med Surg.* 2001;11(1):13-16.

95. Gál J, Csikó G, Pásztor I, Boleskey-Molnár A, Albert M. First description of a papillary carcinoma in the thyroid gland of a red-eared slider (*Trachemys scripta elegans*). *Acta Vet Hung.* 2010;58:69-73.

96. Sykes JM, Trupkiewicz JG. Reptile neoplasia at the Philadelphia Zoological Garden, 1901-2002. *J Zoo Wildl Med.* 2006;37(1):11-19.

97. Mauldin CN, Done LB. Oncology. In: Mader DR, ed. *Reptile Medicine and Surgery*. Philadelphia, PA: WB Saunders;2006:299-322.

98. Bezjian M, Diep AH, de Matos R, Schaefer D. Chinese box turtle (*Cuora flavomarginata*) with lymphoid leukemia characterized by immunohistochemical and cytochemical phenotyping. *Vet Clin Pathol.* 2013;42(3):368-376.

99. Work TM, Balazs GH, Rameyer RA, et al. Retrospective pathology survey of green turtles *Chelonia mydas* with fibropapillomatosis in the Hawaiian Islands, 1993-2003. *Dis Aquat Organ.* 2004;62:163-176.

100. Schumacher JR, Bennett A, Fox LE, et al. Mast cell tumor in an eastern kingsnake (*Lampropeltis getulus getulus*). *J Vet Diagn Invest.* 1998;10:101-104.

101. Schmidt RE, Reavill DR. Metastatic Chondrosarcoma in a corn snake (*Pantherophis guttatus*). *J Herpetol Med Surg.* 2012:22(3-4):67-69.

102. Pereira ME, Viner TC. Oviduct adenocarcinoma in some species of captive snakes. *Vet Pathol.* 2008;45:693-697.

103. Diethelm G, Stauber E, Tillson M, Ridgley S. Tracheal resection and anastomosis for an intratracheal chondroma in a ball python. *J Am Vet Med Assoc.* 1996;209:786-788.

104. Drew ML, Phalen DN, Berridge BR, Johnson TL, et al. Partial tracheal obstruction due to chondromas in ball pythons (*Python regius*). *J Zoo Wildl Med.* 1999;30:151-157.

105. Greenacre CB, Ritchie BW, Latimer KS. Treatment of cartilaginous hyperplasia in ball pythons. *Compend Contin Educ Vet.* 1999;21(7):633-637.

106. Orós J, Tucker S, Fernández L, Jacobson ER. Metastatic squamous cell carcinoma in two loggerhead sea turtles *Caretta caretta*. *Dis Aquat Organ*. 2004;58:245-250.

107. Orós J, P Monagas P, Andrada M, Calabuig P, et al. Metastatic fibrosarcoma in a captive Saharan horned viper (*Cerastes cerastes*) with high hepatic levels of cadmium. *Vet Rec*. 2009;164:690-692.

108. Schmidt RE. Plasma cell tumor in an East Indian water lizard (*Hydrosaurus amboinensis*). *J Wildl Dis*. 1977;13:47-48.

109. Jacobson ER, Ackerman N. What is your diagnosis? Snake swelling. *J Am Vet Med Assoc*. 1981;179(11): 1311-1312.

110. Martin JC, Schelling SH, Pokras MA. Gastric adenocarcinoma in a Florida indigo snake (*Drymarchon corais couperi*). *J Zoo Wildl Med*. 1994;25(1):133-137.

111. Schlumberger HG, Lucke B. Tumors of fishes, amphibians, and reptiles. *Cancer Res*. 1948;8:657-754.

112. Hernandez-Divers SM, Garner MM. Neoplasia of reptile with an emphasis on lizards. *Vet Clin North Am Exot Anim Pract*. 2003;6:251-273.

113. Ritter JM, Garner MM, Chilton JA, Jacobson ER, Kiupel M. Gastric neuroendocrine carcinomas in bearded dragons (*Pogona vitticeps*). *Vet Pathol*. 2009;46:1109-1116.

114. Lyons JA, Newman SJ, Greenacre CB, Dunlap J. A gastric neuroendocrine carcinoma expressing somatostatin in a bearded dragon (*Pogona vitticeps*). *J Vet Diagn Invest*. 2010;22:316-320.

Diagnosis and Treatment of Diseases of the Reptile Cranial Coelom

J. Jill Heatley, DVM, MS, Dipl ABVP (Avian), Dipl ABVP (Reptilian and Amphibian), Dipl ACZM

Session #290

Affiliation: From Zoological Medicine Service, College of Veterinary Medicine and Biomedical Sciences, College Station, TX 774843, USA.

Abstract: The diagnosis and treatment of disease conditions of the reptilian cranial coelom (thorax) is reviewed. By organ, the review includes the thyroid, thymus, trachea, lungs, esophagus, and stomach. This academic review extends well beyond the scope of standard reptilian practice and has the goal of improving reptile practice, and to point out areas for needed research, based on standards and information adapted from that known in human and companion mammals suffering similar diseases. Many reptile diagnostics indicated for disease diagnosis in other species lack basic normative clinicopathologic baseline data required for diagnosis based on hormonal or other analytes. Clinical signs, imaging and histopathologic diagnosis often remain required for definitive diagnoses in reptilian medicine.

Introduction and Overview

This review of the diagnosis and treatment of diseases of the reptile cranial coelom will be based on diseases commonly encountered and diagnostics documented in reptiles and the appropriate therapeutics. Some organs within the cranial coelom, but extending past this region (the trachea, the stomach) will be included and some glands which may or may not be located in the cranial coelom, based on species, will also be included. Diagnostics reviewed will include both those indicated for diagnosis in reptiles based on literature review as well as those successfully used in reptiles. Applicable antemortem diagnostics that have only been reported postmortem will be included. Diagnostics and treatment applicable to the vasculature and heart will not be included in this review.

Trachea

Common diseases of the reptile trachea include trauma, foreign body obstruction, inflammation based on infection (tracheitis) and neoplasia. Uncommon diseases include pulmonary tracheal bronchial prolapse and death in a giant New Caledonian Gecko *(Rhacodactylus leachianus)* and tracheal obstruction caused by fungal granuloma in sea turtles. Many serious viral, bacterial, and fungal diseases (mycoplasmosis, herpesvirus, iridovirus, nidovirus, coronavirus, adenovirus, ferlavirus [OPMV]) have necrotizing tracheitis as a sequella; etiologic agents may proliferate in tracheal mucosa. Based on literature review, tracheitis and obstruction secondary to intubation trauma has not been reported in reptiles unlike in avian species. Vitamin A deficiency also may contribute to tracheitis.

Diagnosis of tracheal disease in reptiles can be challenging based on currently available imaging and other diagnostics. In children, bacterial tracheitis can result in mucopurulent exudates that may acutely obstruct the upper airway and definitive diagnosis remains visualization of the trachea via bronchoscopy.[1] In reptiles, diagnosis of tracheitis, as in other species, generally extends from clinical examination and history. Unlike in companion mammals, because of lack of diaphragm, a cough is not generally appreciated in affected reptiles; however, irritation, caseated tracheal phlegm and dyspnea may be present. Copious discharge may obstruct the tracheal lumen resulting in open mouth breathing and vertical positioning of the head and neck.[2] Affected snakes may extend or lift the cranial portion of the body to facilitate movement of air through the trachea, despite the lumen narrowing based on mucous accumulation. [3,4]

Direct endoscopic examination may be performed, but direct visual guidance of the scope is necessary to avoid damage to tracheal mucosa. Divers reports that complete tracheal occlusion for several minutes is well tolerated[5]; however, preoxygenation would definitely be recommended. Parasites, abscesses, granulomas, trauma, mucosal defects, chondromas and other neoplasms may be seen. Swab samples for cytology, culture and PCR are indicated for lesions consistent with infectious or inflammatory disease. Radiographs may also be supportive in the diagnosis of chondroma in ball pythons (*Python regius*); as CT imaging advances, this modality may become useful for the diagnosis of tracheal disease in reptiles.

Tracheal wash (often erroneously termed transtracheal wash) of reptiles is generally performed via sterile vascular or urinary catheter placed through the glottis. Sedation may or may not be necessary based on the species anatomy and patient temperament for best sample collection. Lidocaine may facilitate glottal relaxation. Methodical care to avoid glottal and oral contamination during catheter placement is necessary. Based on the relatively open lung structures of the reptile, tracheal washes, dependent upon the reptiles positioning, often represent lung washes and procedures for BAL are unnecessary but tracheal wash could be termed bronchofaveolar lavage (BFL).[6] The technique is perhaps best described for the American alligator which was performed endoscopically: Sterile saline is instilled cranial to the tracheal bifurcation, and the animal is slightly tipped ventrally as fluid is aspirated back into a syringe. Calculation of appropriate volumes to instill for reptile tracheal wash appear adapted from mammalian recommendations, and range from 1-10 ml/kg. Samples obtained may be used for PCR, culture and cytology. Bacterial flora and cytology of tracheal wash determined in American alligators consisted of small amounts of mucus, low numbers of ciliated columnar epithelium, cuboidal epithelium, and keratinized squamous epithelial cells in 90% of samples collected.[7] All samples lacked bacteria, parasites and seasonal variation. To my knowledge, specific clinicopathologic blood analyte abnormalities have not been associated with tracheal disease in reptiles; however, serology and PCR for select diseases known to affect the trachea and culture and sensitivity are indicated in cases of suspected viral or bacterial infection.

Treatment of tracheitis/bronchitis should be based on diagnosis and removal of underlying causes. In human pediatric patients suffering viral and or bacterial tracheitis, management includes close observation and monitoring, early initiation of broad spectrum antibiotics, pain management and aggressive airway clearance techniques. Similarly, general treatment recommendations for ferlaviral infection (OPMV), which is associated with purulent tracheal exudate likely based on secondary opportunistic bacterial infection, includes supportive care to include antimicrobial therapy, fluid therapy and nutritional support.[3,4] Systemic treatment may be of less efficacy for treatment of tracheitis than of pneumonia based on the relatively poor blood supply to the trachea; nebulization may be better suited for tracheal delivery of therapeutics. Provision of adequate humidity and hydration and POTZ to facilitate best function of mucociliary apparatus is essential. Expectorants, proteolytics, mucolytics, and bronchodilators (N-acetyl-L cysteine, mucomyst; *guaifenesin*, mucinex, terbutaline, silbutamol) may be of use (Table 1); however, these agents have yet to be specifically assessed for efficacy or safety in reptile patients. Analgesia and/or anti-inflammatory agents deserve consideration in these cases. Infectious disease precautions, as those used for *Salmonella* spp. and infectious agents which may be transmitted via aerosol, should be provided for these patients.

Table 1. Therapeutics for consideration in reptiles with tracheal, lung, esophagus or stomach disease.

Drug	Indication	Dose	Comments and cautions
Acetylcysteine	Mucolytic, thins loosens mucus, bronchitis, pneumonia.	Pediatric[a] nebulized: 1-10 ml of 20% soln, or 2-20 ml of 10% soln, q2-6h.	Very unpleasant taste and odor.
Aminophylline	Acute respiratory disease.	Aminophylline 2.5 mg/ml + sterile saline followed by nebulization with antimicrobials q12h[22]. 50-75 mg/snake 0.7-1.7 kg + 100 ml sterile water nebulize for 30 min q12h[2,9].	Decreases sedative effects of propofol, can lead to theophylline toxicity. Diuretic, central nervous system and cardiac stimulant.
Guaifenesin	Expectorant, thins loosens mucus in airways, clears congestion, cough suppressant, decongestant.	Pediatric[a] immediate release form: 12 mg/kg/day orally, 6 divided doses Sustained release form: 300 mg q12 NTE 600 mg q24h.	Should be administered with fluids in hydrated patient.
Terbutaline	Bronchodilator, bronchospasm prevention.	0.01-0.02 mg/kg IM[3,4] pediatric, acute asthma[a] SC: 0.005-0.01 mg/kg, max dose 0.4 mg, q15-20 min for 2 doses, nebulization: 0.01-0.03 mg/kg min-max dose 0.1-0.25 mg, dilute in 1-2 ml saline q4-6h, continuous IV infusion: 0.08-6 μg/kg/min.	Side effects: anxiety, tachycardia, tremors, nervousness, headache, hyperglycemia, hypo-kalemia, hypotension, pulmonary edema (rare).
Salbutamol Albuterol	Alleviation of bronchospasm, short-acting β_2 adrenergic receptor agonist that relaxes airway smooth muscles.	Pediatric, acute asthma:[3,4] albuterol: 0.05-0.15 mg/kg q4-6h, subsequent doses titrated based on clinical response.	Use with caution based on cardiovascular side effects, hypokalemia.
Orajel PM, Colgate Orabase Soothe-N-Seal, Zilactin, ORA5, Rincinol PRN	Oral ulceration, barrier products, local wash sheet or gel placement, provides pain relief, bioactive barrier.	Topical application PRN[b]	Avoid overdose of benzocaine in small animals.

Drug	Indication	Dose	Comments and cautions
Sodium alginate Sodium polymannuronate Gaviscon® Advance Peppermint SA 100 mg/ml, KHCO3 20 mg/ml	Esophageal ulceration, mucoadhesive polymer sodium alginate with esophageal adhesion properties, bioadhesive upon hydration, capable of self-repair or readherence to the next contact point if dislodged, suspension reacts with gastric acid to form a raft of alginic acid gel of near-neutral pH, floats on stomach contents, impedes gastro-esophageal reflux up to 4 h, protects esophagus from acid, pepsin and bile, in severe cases, raft may be refluxed into esophagus and exert a demulcent effect.	0.1 ml/kg PO PRN up to q6h[15]. Modified adult dose, may not be useful in species without adequate gastric acid secretion for activation.	Use with caution in Na, K, congestive cardiac failure, renal failure or drugs that increase potassium CACO3, hypercalcemia, nephrocalcinosis, recurrent calcific renal calculi.
Sucralfate 100 mg/ml suspension	Oral, gastric or duodenal ulceration.	40-80 mg/kg PO q6-24h,[a] modified pediatric dose.	Renal failure, use with caution, aluminum salt may accumulate.
Viroxyn (benzalkonium chloride)	Herpetic oral ulcers, 0.13 percent benzalkonium chloride, disrupts viral lipid coating and kills on contact.	Single topical application to affected area.[b]	

[a]Suggested, adapted dosages based on drug insert information as found in www.drugs.com accessed June 12, 2016.

[b]Suggested dosages and usage based on http://www.rdhmag.com/articles/print/volume-24/issue-12/feature/treating-oral-ulcers.html.

Tracheal obstruction based on chondroma, cartilaginous granuloma, lymphoma or mucoid tracheal and pulmonary exudate have been treated with tracheal resection and anastomosis and lung cannulation in snakes.[2-4, 8-10] Tracheal resection and anastomosis has been performed in snakes, secondary to removal of obstructive chondroma, and tracheal surgery appears well tolerated.

Esophagus

Compared to domestic mammals, esophageal disease of reptiles appears rare. Congenital anomalies affecting the esophagus, megaesophagus or other esophageal motility disorders appear nonextant. In humans, characteristic symptoms of esophagitis include odynophagia (retrosternal pain on swallowing), dysphagia, nausea and substernal burning pain. Clinical signs may be lacking or could include inanition, regurgitation or, rarely, drooling; however, sea turtles have suffered esophageal impaction with crude oil tar balls and in all aquatic turtles, fish hook and other gear ingestion and lodging within the esophagus is not uncommon.[11] Plaques and erosions in the esophagus commonly occur based on a variety of infectious diseases (ranaviruses adenovirus and others). Based on reported pathology findings, viral particles or inclusions bodies could be found based on swab or biopsy samples, respectively which could be assessed by cytology, culture and PCR. As a widely distensible potential space, indirect imaging may have limited diagnostic capability in the palpably normal esophagus. Endoscopy with gas or liquid insufflation is indicated for lesion visualization and sampling; Intubation is required and downward patient positioning is recommended to avoid aspiration pneumonia. As in humans, no hematologic or biochemical analyte abnormalities are specific to the diagnosis of esophageal disease; however, in reptiles with persistent regurgitation, electrolyte alteration would be a supportive finding.[12]

In humans esophageal infection is rare and associated with immunocompromise. Within this population, Herpes simplex, cytomegaloviral and candidal infections and idiopathic esophageal ulceration are common. Recent human studies of these populations provide data relevant to reptiles (especially tortoises) suffering esophageal ulceration. Human symptoms characteristic of esophagitis include odynophagia (retrosternal pain on swallowing), dysphagia, nausea and substernal burning pain.[13] In a prospective endoscopic study of these lesions, brush samples (for cytology or culture) were more sensitive for diagnosis of candidal or viral infection than biopsy for histology. Blind brushing of the esophagus via esophageal tube has a reasonable sensitivity for each disease at 84 and 75%, respectively.[14] Treatment of esophageal disease is aimed at etiology identification and systemic treatment; however, systemic analgesics and liquid topical antiulcer medications (sucralfate and sodium alginate) also deserve consideration in these cases.[15]

Lungs

Lung disease is commonly recognized in reptilian species based on infectious and traumatic causes, although disseminated or primary neoplasia and nutritional disease (hypovitaminosis A) may also affect the lungs. Acute pulmonary edema, chronic obstructive pulmonary disease and asthma remain unreported in reptiles; however, severe obesity can occur in reptiles and could result in similar signs (Pickwickian tortoise). Clinical signs may be lacking or may include obvious damage based on trauma, increased respiratory rate or effort, abnormal respiratory sounds, or, in aquatic species, abnormal buoyancy. In chelonians, the author finds percussion more useful than auscultation. The challenging diagnosis and occult nature of pneumonia in reptiles is similar to the disease in geriatric humans, in which arterial blood gas and lactate are often used to facilitate diagnosis.[16]

Imaging may be accomplished via direct or indirect methods. Horizontal beam radiographs 3 or 2 view as dictated by species are indicated for diagnosis of respiratory disease; however, CT and MRI are more sensitive although also more expensive for detection of respiratory tree abnormalities and normal snake lung images have been determined.[17-20] Approaches for endoscopic examination and biopsy of the lung generally requires pulmonoscopy; the approach is determined based on species anatomy and lesion localization from indirect imaging. Air sac pulmonoscopy of snakes or a carapacial or prefemoral lung assessment approach in chelonia has been described. Sample collection via biopsy, wash and brush may be performed for evaluation of cytology, histology, and culture and PCR. Blood gases and hemoximetry may be useful to diagnose respiratory acidosis, oxygenation status, and acid base status, should normal values be available from the same species and preferably the same analyzer or monitor. These analytes would likely prove most useful in evaluating a patient's response to therapeutic interventions and for monitoring severity and progression of disease.[21] Leukocytosis is likely to occur in patients with pneumonia of infectious etiology; however, one would expect a more delayed response in the leukogram than in blood gases when assessing disease progression or patient response to intervention.

Treatment of pneumonia and other inflammatory or infectious diseases of the lung of reptiles may have a prolonged course prior to resolution. Viral, parasitic, bacterial or fungal etiologies are expected etiologies. A combination of nebulization and systemic treatment as well as excellent supportive care are recommended. In the author's experience, lung biopsy of chelonians may result in lateral coelomic skin extension "puffing out" with slow resolution not to be confused with actual coelomitis. Similarly, the placement of pneumonic catheters for direct pulmonary therapy is not without potentially life threatening consequence.

Stomach

Common diseases of the reptile stomach include foreign body ingestion/ impaction/ bloat, gastric ulceration (green iguanas, *Iguana iguana*) cryptosporidial infection, neoplasia (gastric adenoma of bearded dragons, *Pogona* spp.), coccidiosis, viral, fungal and bacterial infections and idiopathic atrophic gastritis. Clinical signs of gastric disease in reptiles may include anemia, lethargy, anorexia, vomiting and melena.[23,24]

Diagnosis of gastric disease via plain radiography in most reptiles is challenging based on lack of normal values for gastric positioning and size, lack of coelomic fat to outline internal organs and the natural position of most organs within the coelom which lie in a similar plane to the stomach. Contrast radiography (barium sulfate or iodinated contrast media may be used) may be helpful, but GI transit time may be prolonged in many of these species.[23] CT or MRI combined with intravenous administration of iodinated contrast media are recommended to provide greatly enhance detail for these structures; For chelonians CT scans repeated at 1, 3 and 10 minutes after contrast injection may improve the diagnostic quality of the images. In CT images, the stomach is easy to recognize when gas is present, but can otherwise be difficult to differentiate from surrounding soft tissue unless oral contrast media is administered.[23]

Ultrasound is likely best used when the GI tract is fluid filled or empty and may be less useful in the herbivorous reptile based on gas production. Thus carnivorous reptiles suffering ileus or obstruction may be the best clinical choice to obtain ultrasound diagnosis; however, the stomach was well evaluated via ultrasound in a study of young green iguanas.[23] In the ball python, a left approach revealed the stomach with the pylorus identified medial and dorsal to the gall bladder.[25]

As in evaluation of the esophagus, stomach endoscopy is performed with insufflation of air or saline, and facilitates gross evaluation of lesions, foreign body retrieval and obtaining samples for cytology, histology, culture, microscopic observation of parasites and PCR. In contrast to the delicate esophagus of some species, especially those suffering erosions and necrotic lesions, the stomach is generally thicker walled and more amenable to biopsy. Stomach wash via blind placement of a red rubber catheter may be a less invasive and expensive way to obtain wash samples for similar testing, but does not allow for histology, object retrieval or direct observation of gastric lesions.

Normal gastrointestinal flora is not well characterized in most reptiles. Unlike in mammalian species, specific clinicopathologic plasma biomarkers for gastrointestinal function remain lacking in reptiles. Anemia of chronic disease or from gastrointestinal blood loss and/or electrolyte abnormalities may be present, but the author finds a lack of clinicopathologic abnormalities to be the most common finding in these cases, followed by leukopenia and anemia.[12]

Thyroid

In reptiles, the thyroid regulates ecdysis, reproduction, tail regeneration, growth, endocrine function, and metabolic rate.[26] The role of external factors in thyroid function of reptiles including light cycles, hibernation, and thermal gradients must be considered in suspect cases.[23] Diseases of the thyroid appear not uncommon in reptiles and if neoplasia is included, abnormal thyroid glands have been reported in all classes of reptiles. As in cats, lizard antemortem diagnosis may be based on clinical signs and increased total T4.[27] Clinical signs reported in lizards include those linked to the mass itself: Intraoral mass, ventral neck swelling, palpably enlarged thyroid gland, oral hemorrhage and behavioral changes associated with hormonal changes such as polyphagia, aggression, weight loss, tachycardia. Increased ecdysis with response to methimazole treatment prompted the diagnosis of hyperthyroidism in an older corn snake, but thyroxine concentration were not determined.[28] Total T4 Values have been reported in a number of other reptilian species but vary seasonally in some species and relatively low "normal" values may make diagnosis challenging via hormonal assay.[26] Although rare, The likelihood of metastasis in cases of thyroid carcinoma dictates that radiographs or advanced imaging be a part of the diagnostic work up of these cases.

Goiter (enlarged thyroid) has been reported to affect tortoises on iodine deficient diets or diets with dietary goitrogens: therefore these should be eliminated or minimized in the herbivorous reptile diet: bok choy, broccoli, cabbage, cauliflower, kale, mustard seed, rapeseed, soy bean sprouts, and turnips.[26] Oxalate crystals were found in free-living desert tortoises with thyroid enlargement but without other apparent disease consequence.[29] The diagnosis of hypothyroidism in reptiles via thyroid hormones and TSH remains poorly validated, but the disease appears rare and controversy exists regarding differentiation of obesity and hypothyroidism in tortoises.

Thymus

True thymic disease appears rare in reptiles, with reports limited to a single case in young Galapagos tortoises (*Chelonoidis niger*).[30] Rule outs of expected seasonal change, or generalized lymphatic proliferation from infectious agents or other inflammatory diseases should be considered in these cases. In humans, tumors of the thymus (thymomas and thymic carcinomas) are rare intrathoracic tumors of a variety of cell type origins; a correlation with autoimmune disorders exists in about 1/3 of human cases (Myasthesia gravis, red cell aplasia and hypogammaglobulinemia). Thoracic CT imaging is considered similar or superior to MRI in these cases. In humans, thymoma is more likely to be associated with autoimmune disease while thymic carcinoma cases typically have nonspecific local symptoms. Thus clinicopathologic workup in humans includes complete blood cell count, reticulocyte count and serum protein electrophoresis, as well as anti-acetylcholine receptor and anti-nuclear antibody tests.[31] Relevant differential diagnoses include lymphomas, germ-cell tumours (teratoma or seminoma/non-seminomatous tumours),and primary lung carcinoma.[31] In reptiles, biopsy is recommended prior to resection or other treatment and remains the best option for definitive diagnosis, but may still result in equivocal results.

Parathyroid Glands

While nutritional and renal secondary hyperparathyroidism (NSHP, RSHP) remain common diagnoses in reptilian medicine, primary parathyroid gland disorders appear rare and remain undescribed antemortem likely based on lack of validated laboratory assay for PTH and few normative values for ionized calcium in apparently healthy reptiles.[26] Treatment of NSHP and RSHP is common and beyond the scope of this manuscript. In humans, primary hyperparathyroidism is extremely rare and generally found based on finding hypercalcemia during routine blood screen or in elderly patients with confusion, and dehydration. Of human patients with hypercalcemia on blood testing, approximately 55% are then diagnosed with primary hyperparathyroidism (PHP), based on radioimmunoassay.[32]

Ultimobranchial Glands

Diagnosis and treatment of primary disease arising from the ultimobranchial bodies of reptiles remains undescribed.

References

Trachea

1. Graf J, Stein F. Tracheitis in pediatric patients. *Semin Pediatr Infect Dis*. 2006;17(1):11-13.

2. Myers DA, Wellehan JF, Isaza R. Saccular lung cannulation in a ball python (*Python regius*) to treat a tracheal obstruction. *J Zoo Wildl Med*. 2009;40(1):214-216.

3. Schumacher J. Respiratory diseases of reptiles. *Semin Avian Exot Pet Med*. 1997;6(4):209-215.

4. Schumacher J. Reptile respiratory medicine. *Vet Clin North Am Exot Anim Pract*. 2003;6(1):213-231.

5. Divers SJ. Reptile diagnostic endoscopy and endosurgery. *Vet Clin North Am Exot Anim Pract*. 2010;13(2):217-242.

6. De la Navarre BJ. Common procedures in reptiles and amphibians. *Vet Clin North Am Exot Anim Pract*. 2006;9(2):237-267.

7. Lafortune M, Göbel T, Jacobson E, Heard D, Brown D, Alleman R, Vliet K, Harr KE, Hernandez J. Respiratory bronchoscopy of subadult American alligators (*Alligator mississippiensis*) and tracheal wash evaluation. *J Zoo Wildl Med*. 2005;36(1):12-20.

8. Diethelm G, Stauber E, Tillson M, Ridgley S. Tracheal resection and anastomosis for an intratracheal chondroma in a ball python. *J Am Vet Med Assoc*. 1996;209(4):786-788.

9. Drew ML, Phalen DN, Berridge BR, Johnson TL, Bouley D, Weeks BR, Miller HA, Walker MA. Partial tracheal obstruction due to chondromas in ball pythons (*Python regius*). *J Zoo Wildl Med*. 1999;1:151-157.

10. Summa NM, Guzman DS, Hawkins MG, Grosset C, Chen VS, Goldsmith D, Keel K, Woolard K, Young AC, Bucy DS, Steffey MA. Tracheal and colonic resection and anastomosis in a boa constrictor (*Boa constrictor*) with T-cell lymphoma. *J Herpetol Med Surg*. 2015;25(3-4):87-99.

Esophagus

11. Orós J, Torrent A, Calabuig P, Déniz S. Diseases and causes of mortality among sea turtles stranded in the Canary Islands, Spain (1998-2001). *Dis Aquat Org*. 2005;63(1):13-24.

12. Mitchell MA, Diaz-Figueroa O. Clinical reptile gastroenterology. *Vet Clin North Am Exot Anim Pract.* 2005;8(2):277-298.

13. Baehr PH, McDonald GB. Esophageal infections: risk factors, presentation, diagnosis, and treatment. *Gastroenterol.* 1994;106(2):509-532.

14. Bonacini M, Young T, Laine L. The causes of esophageal symptoms in human immunodeficiency virus infection: a prospective study of 110 patients. *Arch Intern Med.* 1991;151(8):1567-1572.

15. Smart JD, Dunkley S, Tsibouklis J, Young S. An evaluation of the adhesion of solid oral dosage form coatings to the oesophagus. *Int J Pharm.* 2015;496(2):299-303.

Lungs

16. Wunderink RG, Waterer GW. Community-acquired pneumonia. *N Engl J Med.* 2014;370(6):543-551.

17. Pees M, Kiefer I, Thielebein J, Oechtering G, Krautwald-Junghanns ME. Computed tomography of the lung of healthy snakes of the species *Python regius*, *Boa constrictor*, *Python reticulatus*, *Morelia viridis*, *Epicrates cenchria*, and *Morelia spilota*. *Vet Radiol Ultrasound.* 2009;50(5):487-491.

18. Pees MC, Kiefer I, Ludewig EW, Schumacher JP, Krautwald-Junghanns ME, Oechtering GU. Computed tomography of the lungs of Indian pythons (*Python molurus*). *Am J Vet Res.* 2007;68(4):428-434.

19. Pees M, Kiefer I, Oechtering G, Krautwald-Junghanns ME. Computed tomography for the diagnosis and treatment monitoring of bacterial pneumonia in Indian pythons (*Python molurus*). *Vet Rec.* 2008;163(5).

20. Pees M, Kiefer I, Ludewig E, Westphal S, Schmidt V, Krautwald-Junghanns ME, Oechtering G. Comparative use of modern imaging techniques for the diagnosis of pneumonia in three Indian pythons (*Python morulus*). *Tierärztliche Praxis Kleintiere.* 2006;34(4):275-282.

21. Davis MD, Walsh BK, Sittig SE, Restrepo RD. AARC clinical practice guideline: blood gas analysis and hemoximetry: 2013. *Respir Care.* 2013;58(10):1694-1703.

22. Raiti P. Administration of aerosolized antibiotics to reptiles. *Proc Annu Conf Assoc Rept Amph Vet.* 2002; 119-123.

Stomach

23. Mans C. Clinical update on diagnosis and management of disorders of the digestive system of reptiles. *J Exot Pet Med.* 2013;22(2):141-162.

24. Kubisch U, Fischer I, Hatt JM. Gastric ulcer in green iguanas (*Iguana iguana*). *Tierärztliche Praxis Kleintiere.* 2006;34(1):50-53.

25. Banzato T, Russo E, Finotti L, Milan MC, Gianesella M, Zotti A. Ultrasonographic anatomy of the coelomic organs of boid snakes (*Boa constrictor imperator*, *Python regius*, *Python molurus molurus*, and *Python curtus*). *Am J Vet Res.* 2012;73(5):634-645.

Thyroid

26. Rivera S, Lock B. The reptilian thyroid and parathyroid glands. *Vet Clin North Am Exot Anim Pract*. 2008;11(1):163-175.

27. Hernandez-Divers SJ, Knott CD, MacDonald J. Diagnosis and surgical treatment of thyroid adenoma-induced hyperthyroidism in a green iguana (*Iguana iguana*). *J Zoo Wildl Med*. 2001;32(4):465-475.

28. Frye FL. Biomedical and Surgical Aspects of Captive Reptile Husbandry. Malabar, FL: Krieger Publishing, Co.;1991.

29. Jacobson ER, Berry KH, Stacy B, Huzella LM, Kalasinsky VF, Fleetwood ML, Mense MG. Oxalosis in wild desert tortoises, *Gopherus agassizii*. *J Wildl Dis*. 2009;45(4):982-988.

Thymus

30. Fleming GJ, Heard DJ, Uhl EW, Johnson CM. Thymic hyperplasia in subadult Galapagos tortoises, *Geochelone nigra*. *J Herpetol Med Surg*. 2004;14(1):24-27.

31. Girard N, Ruffini E, Marx A, Faivre-Finn C, Peters S. Thymic epithelial tumours: ESMO clinical practice guidelines for diagnosis, treatment and follow-up. *Ann Oncol*. 2015;26(suppl 5):v40-55.

Parathyroid

32. Mundy G, Cove D, Fisken R. Primary hyperparathyroidism: changes in the pattern of clinical presentation. *Lancet*. 1980;315(8182):1317-1320.

Chelonian Conservation-Breeding: Health and Husbandry

Paul M. Gibbons, DVM, MS, Dipl ABVP (Reptile and Amphibian)

Session #325

Affiliation: From the Turtle Conservancy, 1794 McNell Rd, Ojai, CA 93023, USA

Abstract: Reptile health depends upon good husbandry, and the best captive care is based upon knowledge about each species' individual native microhabitat and seasonality. This masterclass will describe health and husbandry challenges and successes over the 10-year history of a large colony of threatened tortoise and freshwater turtle species at an Association of Zoos and Aquariums-certified facility. Descriptive data are presented at the population level as well as individual unique cases and outbreaks.

History

In 2004, John L. Behler (1943-2006), then curator of reptiles at the Bronx Zoo in New York City, sought out conservationist Eric Goode to ask whether he would be willing to provide a permanent home for the zoo's long-standing collection of endangered tortoises being kept at St. Catherine's Island in Georgia. Goode, a lifetime turtle enthusiast, rose to the challenge and transformed his Southern California retreat into the Turtle Conservancy, a non-profit facility certified by the Association of Zoos and Aquariums (AZA) and dedicated to protecting the most endangered turtles and tortoises and their habitats worldwide. In 2006, we named the new campus the Behler Chelonian Center to honor John's life of dedicated service to turtle and tortoise conservation.

Since that time, the Turtle Conservancy has become one of the most successful conservation-breeding centers in the world. We have hatched over 1,500 offspring from 21 different species, 99% of which are threatened with extinction, according to the Tortoise and Freshwater Turtle Specialist Group of the IUCN. Eighteen of the 44 species we have cared for over the years are critically endangered, 13 are endangered and 2 are vulnerable. Experienced turtle keepers would agree that several of these species are difficult to keep alive, let alone breed in captivity. Examples include the impressed tortoise (*Manouria impressa*), forest hinge-back tortoise (*Kinixys erosa*), Chaco tortoise (*Chelonoidis chilensis*), Okinawa leaf turtle (*Geoemyda japonica*), Madagascar flat-tailed tortoise (*Pyxis planicauda*) and all three subspecies of spider tortoise (*Pyxis arachnoides*). Such delicate species require a peaceful existence with minimal human contact to thrive, and we can focus on husbandry at the Turtle Conservancy, instead of attendance and entertainment like most zoos, because we are not open to the public.

Captive-Bred Animals

The Turtle Conservancy takes its mission seriously, protecting species from the threats that drive them toward extinction. Our animals come from other zoological institutions, wildlife authorities that have confiscated illegally trafficked animals around the world, and from private individuals who donate their animals for conservation. Our policy is to not buy wild-caught animals; therefore, we do not contribute to one of the greatest threats to endangered turtles and tortoises. We strongly encourage hobbyists and pet owners to demand captive-bred reptiles instead of supporting the global wildlife trade.

The number of pet turtles in the United States is growing, and commercial trade in wild-caught reptiles is one of the factors driving many species toward extinction. Alternatively, when you choose to purchase captive-born reptiles from local or regional breeders instead of buying wild-caught animals, you not only reduce pressure on wild populations, but if you breed them, you also might help prevent their extinction. The African spurred tortoise (*Centrochelys sulcata*) is just one example of how an endangered species can become one of the most common pet reptiles and be at lower risk of extinction because of the work of hobbyists and private breeders. Many other species might also be helped in this way, and the Turtle Conservancy is setting an example for the zoological community by offering a limited number of our captive-bred offspring to qualified private individuals for this very reason. All proceeds from these adoptions directly support our in range global conservation programs. Together, we can ensure these species live on.

International Conservation

The offspring we hatch at the Turtle Conservancy participate in programs that help save their species from extinction. Accordingly, the best possible outcome is to repatriate our captive-born offspring to their native habitats. This is more difficult than one might imagine, and amazingly, with the five yearling golden coin turtles (*Cuora trifasciata*) we sent to Hong Kong for a reintroduction program in January 2013, the Turtle Conservancy became the first organization to return foreign-born, captive-bred turtles to their native country for conservation.

We also share offspring with other conservation centers to create multiple assurance colonies around the world, providing a safety net in case of natural disasters, political uprisings or insufficient funding. For example, in May 2014 we sent Burmese star (*Geochelone platynota*) and Burmese black mountain tortoises (*Manouria emys phayrei*) to initiate colonies at Richard Branson's conservation center on Necker Island in the British Virgin Islands. In addition to these programs, a few of our offspring will remain in AZA institutions to satisfy the needs of their Species Survival Plan programs and ensure genetic variability in the professionally managed captive populations.

The Turtle Conservancy's in-country conservation programs include habitat protection and stewardship, turtle and tortoise reintroduction and education, curbing the illegal wildlife trade, and a seed grant program. Our work currently spans the globe and includes 14 countries. For example, we are protecting one of the last strongholds of the Critically Endangered geometric tortoise (*Psammobates geometricus*) in South Africa with an assemblage of adjacent tracts of intact fynbos ecosystem that now exceeds 800 acres and contains the largest known population of this species. We work with partners to protect the Critically Endangered ploughshare tortoise (*Astrochelys yniphora*) in Madagascar by engaging in community education programs and by marking the shells of wild tortoises with an indelible inscription to thwart poachers.

You Can Help

The Turtle Conservancy's work protecting species from extinction can only be sustained through generous public support, and even small donations help feed the animals and keep them healthy. Our goals are not just to be a last line of defense against extinction, like Noah's Ark, but also to actively refine husbandry techniques, enhance reproduction and protect native habitats to ensure that wild populations are self-sustaining.

What can you do? First, know the laws and report illegal animal trafficking activities to the authorities, especially the U.S. Fish and Wildlife Service. Second, buy reptiles that were captive-born in your region at facilities such as The Turtle Conservancy. Third, volunteer with local, regional and international programs that help protect species from extinction. And fourth, donate generously to organizations that protect threatened species. Learn more about our work at TurtleConservancy.org and follow us on Twitter @turtletweets

On-Site Greenhouses

The Turtle Conservancy's formula for husbandry success relies heavily on knowledge about each species' unique natural history. We personally undertake field expeditions to the native range of species under our care to fill in the gaps in available scientific evidence regarding seasonal and daily variations in temperature, humidity, rainfall, fog, water sources, canopy, diet, shelter, substrates and activity. At our conservation-breeding center, we continually define and redefine the best possible husbandry techniques to provide species-specific microhabitat needs with a limited number of indoor environments and the Mediterranean climate of California's central coast. We provide as much outdoor time as possible with fresh air, sunshine and natural browse. Although we avoid mixing multiple species in a single enclosure, we do group species from similar climates together in the same building.

We have 2 greenhouses that provide excellent exposure to natural light cycles, but which can be challenging to heat and cool. One simulates a cloud forest and is kept very humid with moderate temperatures, 29°C (84°F) during the day, 22°C (72°F) at night, and the other simulates a lowland hot climate with moderate humidity and a broad diurnal temperature fluctuation, 32°C (90°F) day/20°C (68°F) night. Both employ multiple heating, cooling and moisture systems coordinated by an advanced climate control computer manufactured for commercial greenhouses.

The cloud forest environment primarily houses impressed tortoises and Okinawa leaf turtles. The hot lowland environment houses an array of species, including Sulawesi forest turtles (*Leucocephalon yuwonoi*), Forsten's tortoises (*Indotestudo forstenii*), spiny hill turtles (*Heosemys spinosa*), Arakan forest turtles (*Heosemys depressa*), Oaxaca wood turtles (*Rhinoclemmys rubida rubida*) and Burmese star tortoises, which spend much of their time outdoors.

Background heating systems include circulating warm water in the concrete floor and hot water radiators on the walls. Automated ridge vents open with cooling needs, and when temperatures really heat up, they close almost completely so fresh cool air can freely enter from the air conditioner in the cloud forest greenhouse or the commercial evaporative cooler in the lowland hot greenhouse. Horizontal shade-cloth curtains draw closed to reduce excess radiant heat on sunny days and at night to hold warmth inside. Humidity is maintained with commercial ultrasonic and centrifugal humidifiers.

We provide locally high humidity, condensation and artificial rain inside enclosures using irrigation or fine-mist systems to meet species-specific needs. Basking spots are provided using ultraviolet-B (UVB)-emitting mercury vapor lamps on cloudy days and in enclosures with insufficient sunlight. Some of the greenhouse panels are Acrylite OP-4 acrylic (CYRO Industries, Parsippany, NJ), which transmits some UVB radiation, and others are either Plexiglas or tempered glass, which both block nearly all UVB. We provide compact fluorescent UVB-emitting bulbs for non-basking species that are generally more active at dusk and for those from dense forests with little direct sunlight. Lights are timed to meet the individual needs of each species, including seasonality and basking behavior. The intensity of UVB emitted by each lamp is tested every six months using a meter with peak response at 290 nm, and lamps are replaced when UVB emission falls below 10 microWatts/cm^2 at the basking site.

Other Optimal Environments

In addition to the greenhouses, we constructed two semi-enclosed environments that we call cold frames, with removable translucent panels that constrain temperature extremes and manage humidity. The first of these simulates coastal dry zones with 90% relative humidity at night and 20% during the day. Species housed here include the Egyptian tortoise (*Testudo kleinmanni*), spider tortoise and ploughshare tortoise, which all retreat beneath shrubs or bushes at night and during the hot, dry hours of the day to take advantage of focal moderate humidity.

The second cold frame simulates southern hemisphere scrub habitats with generally cooler temperatures and low to moderate humidity. The species housed here include Chaco tortoises, parrot-beaked tortoises (*Homopus areolatus*) and angulated tortoises (*Chersina angulata*), which all retreat into subterranean burrows at night and during the heat of the day to find moderate humidity and temperature.

We also utilize two well-heated, well-insulated and biologically isolated houses to provide safe environments for hatchlings, animals under quarantine and the most temperature-sensitive species during the winter. These houses employ insulated wood-frame construction with screened windows and skylights that provide natural ambient light cycles and insect-proof ventilation.

In addition, there are four large tortoise houses for hardy species such as radiated tortoises (*Astrochelys radiata*) and Galapagos tortoises (*Chelonoidis nigra* complex). Circulating warm water in the floors and radiators is used for background heating, and basking sites are illuminated as needed. Materials and instruments such as pens, notebooks, scales, hoses, nozzles, scrub brushes, tools, mops, buckets and watering cans are dedicated to each house, and footbaths are used at all entrances and exits. Disposable, biodegradable nitrile gloves are used on individual enclosures, not just in these houses but throughout the center as part of the Turtle Conservancy's comprehensive biosecurity plan.

Ponds, pools and aquariums are managed with recirculating biological filtration with UVB sterilizers in the larger systems. One pond is approximately 110 cm deep and allows for outdoor hibernation of Pan's box turtle (*Cuora pani*). We monitor water quality parameters with a multi-test kit and perform partial water changes as often as twice a week in systems with numerous turtles.

The Casitas

Twelve small houses called casitas provide hardy outdoor species, such as radiated tortoises and Burmese black mountain tortoises, with a warm, dry space on cool nights and during brief periods of winter rain. These small houses are approximately 120 cm long, 90 cm wide and 75 cm tall. They are insulated with R-10 rigid foam insulation beneath the concrete slab and between layers of plywood on the walls, top and doors. Cracks are sealed with weather-stripping to prevent drafts, heat is provided by thermostat-controlled ceramic heat emitters, and daytime light is provided by timer-controlled compact fluorescent bulbs. Interior height is designed so electrical fixtures are high enough to protect the tortoises from thermal burns.

Bedding in the casitas consists of grass hay and rice hulls for radiated tortoises, and sphagnum moss for Burmese black mountain tortoises. We open casita doors in the morning and encourage the tortoises to exit the houses within a few hours. Flexible, translucent vinyl door strips are hung inside the doorframes to retain heat when a door is open on relatively cool days, and most tortoises pass through them readily.

Hatchlings and Juveniles

Six floor-to-ceiling incubators provide enough space for more than 500 eggs at one time. We incubate 130-150 clutches and hatch out 250-300 turtles and tortoises per year. Each incubator is set for temperatures that meet the needs of the species being incubated. High temperature incubators (29°C to 32°C) have a thermostat that allows them to cool down to 24-25.5°C at night, which simulates the daily fluctuations found in natural nests. Incubators for forest and temperate species are kept at relatively cooler temperatures (25-27.5°C) without daily fluctuations because the microhabitats in these regions have more constant diurnal temperatures.

Incubation protocols are based on the natural history of each individual species. Over time we refine them according to our successes and failures. For example, we started incubating parrot-beaked tortoise eggs in their natural nest in the substrate of the enclosure, and we hatched two of six eggs. Then we incubated the next eight eggs in an incubator at either 27.8°C or 30°C because other people reported success at those temperatures, but none hatched. With that new knowledge, we began incubating at 31.7°C; since then seven out of seven parrot-beaked tortoise eggs have hatched. The next step is to define what temperatures produce each sex.

We house juvenile tortoises in tables constructed from common lumber and marine-grade plywood. A table approximately 70 cm wide and 150 cm long can accommodate up to 30 hatchling tortoises with plates, dishes, plants and décor for hiding, browsing and environmental enrichment. We drill holes in the bottom for drainage and paint interior surfaces with floor epoxy to delay moisture damage. A hinged top of hardware cloth excludes predators, especially rats.

A table depth of 35 cm can accommodate a few inches of gravel, a layer of weed cloth and sufficient topsoil substrate for plants to root and tortoises to dig. We also provide a layer of rice hulls mixed with sand to allow juvenile Burmese star tortoises and spider tortoises to bury themselves. At night we cover outdoor tables with thermostat-controlled heating pads and insulated blankets whenever temperatures are predicted to be less than 18°C (65°F). We water the plants once or twice a week, and use a fine mist to cool and moisturize the air on very hot, dry days. We strategically place palm fronds to provide additional patches of shade on sunny days when ambient temperatures exceed 32°C (90°F) because the substrate in direct sunlight can reach in excess of 60°C (140°F).

Keeping records is a major part of managing such a large collection. Each specimen is photographed regularly because individual markings change with age. No identification system is perfect, so at the Turtle Conservancy we use redundant systems including radio frequency identification (aka PIT tags), marginal marks, color coding and painted numbers. Information is recorded in the International Species Information System database using the online software known as Zoological Information Management System. Each species is managed to meet reproductive and demographic needs of its conservation programs. Thefts have occurred at conservation breeding centers around the world, so we maintain a multi-layered security system including infrared motion detectors and video cameras with 24/7/365 monitoring.

Health Concerns in a Large Collection

The individual circumstances that allow pathogens to cause disease differ for each organism, and result from interactions among the environment, the pathogen, and the host. The study of these factors is a field known as "disease ecology." Species-appropriate environmental conditions are paramount for a healthy immune system in reptiles. An "emerging disease" has either appeared recently in a population, is rapidly increasing in incidence, changing in geographic range, or threatens to increase in the near future. Intranuclear coccidiosis, cryptosporidium, ranavirus, and adenovirus are emerging infectious diseases of chelonians. Mycoplasmosis, herpesvirus, and systemic amoebiasis, though no longer emerging, do continue to be important causes of disease in collections. Other parasites include an array of nematodes, flagellated protozoa, and rarely, flukes (*Falcaustra* sp.) and pentastomids. In addition to infectious diseases, several non-infectious problems occur with some frequency. Necropsy is performed on every animal that dies at the Turtle Conservancy and the etiologic categories are shown for 473 mortalities with 70 histopathologic examinations in Table 1.

Table 1. Cause of death for 473 animals at the Turtle Conservancy, 2005-2015.

Etiologic category	Count
Parasites	177
Husbandry related	114
Developmental	34
Bacterial	13
Iatrogenic	8
Metabolic	8
Predator attack	6
Toxic	6
Unknown	106

Preventive medicine

Intestinal parasites can become pathogenic in captive animals because, even though many of the parasite species are not harmful in the wild, captive conditions increase the transmission rate, particularly for species transmitted via the fecal-oral route. Preventive measures are not intended to eliminate nematodes or protozoa, but to keep the intestinal load below the threshold of disease. The antiparasitic agents used most commonly at the Turtle Conservancy include benzimidazoles, metronidazole, and praziquantel.

Intranuclear coccidiosis of testudines

Intranuclear coccidiosis of Testudines (TINC) was first reported over 20 years ago and has been increasingly reported in numerous species in Europe, North America, and possibly Asia. It may be the most important disease currently affecting chelonians, and should be considered for every terrestrial chelonian with systemic illness or clinical signs involving multiple organ systems. Early diagnosis and treatment is essential, and PCR performed on conjunctival, oral/choanal mucosa, and cloacal tissue appears to be the most useful antemortem diagnostic tool. Effective treatment includes providing optimal environmental conditions and either ponazuril or toltrazuril 15 mg/kg PO q48h × 30 days. This prolonged treatment is necessary because these drugs have no effect on oocytes.

Cryptosporidiosis

Cryptosporidium-related disease has recently been described in a number of chelonians globally. Infection is sometimes associated with chronic diarrhea, anorexia, pica, decreased growth rate, weight loss, lethargy, or passing undigested feed. A consensus PCR followed by sequencing can be performed on feces to identify the species of *Cryptosporidium*. No treatments have been shown to clear infection, but paromomycin did eliminate clinical signs of disease in a group of *Testudo hermanni*. Concurrent infection with other pathogens may enhance disease progression.

Ranavirus

Ranaviral infection recently caused mass mortality events in wild North American box turtles (*Terrapene* spp.) and has been found in some captive chelonians that are allowed outdoors. Although the virus has been shown to be widespread in stable, healthy populations, when fulminant infection occurs it is often fatal. Clinical signs are similar to many other systemic diseases, and can include signs of upper respiratory disease, oral ulceration, cutaneous abcessation, subcutaneous edema, anorexia, and lethargy. Ranaviral infection is not acquired orally, and disease may require stress coupled with a vector or open wound. Clinically ill chelonians can be tested for ranavirus via PCR of oral and cloacal swabs, or whole blood.

Adenoviruses

Since 2009, adenoviruses have been identified as the cause of several mortality events in captive chelonians. Disease has involved liver, gastrointestinal tract, spleen, and brain. It affects young or immunocompromised animals such as the recent group of 105 illegally imported Sulawesi tortoises that all died with systemic *Siadenovirus* infection despite aggressive supportive care at several zoological institutions. PCR of cloacal swabs and plasma samples have shown promise for antemortem detection, although treatment has proven unsuccessful to date.

Mycoplasmosis

Upper respiratory tract disease (URTD) can be caused by a number of bacteria and viruses in chelonians. *Mycoplasma agassizii* was first described in 1995, and *M testudineum* was described in 2004. Both organism, as well as other species of *Mycoplasma* can be found in both healthy and ill tortoises. Clinical signs of disease appear with stress (e.g., territorial encroachment, habitat perturbation, relocation, capture) or poor environmental conditions (e.g., insufficient heat, poor nutrition, poor ventilation, overcrowding). Clinical signs of disease include nasal discharge, blepharal edema, and blepharoconjuctivitis observed as bubbly or wet nose, puffy eyelids, and reddened conjunctiva. Debris adhering to the rostrum is a common sign of nasal discharge. Pet tortoises are often presented for decreased appetite, most likely resulting from reduced sense of smell. Affected tortoises often wipe their rostrum with their forearms, which may be covered with wet or dry mucus, often with adherent debris or substrate. These clinical signs are not specific, and the list of differential diagnoses include foreign material (nasal/conjunctival), inhaled/aerosolized irritants, hypovitaminosis A, neoplasia, herpesvirus, ranavirus, opportunistic bacteria, and *Pasteurella testudinis*. Newly infected individuals may not exhibit clinical signs (prepatent period) for up to one year, and signs can persist for months to years. Mycoplasmal organisms are found in healthy tortoises, and many species rarely exhibit clinical signs of illness. In addition to North American tortoises, *Gopherus* spp., the most susceptible species include star tortoises (*Geochelone elegans* and *Geochelone platynota*), Chaco tortoises, *Indotestudo* spp., *Testudo* spp., and Russian tortoises (*Agrionemys horsfieldii*). Serology is available for some species and can help identify newly infected individuals, but chronically infected tortoises can be seronegative. Consensus PCR is the most useful test in a tortoise with clinical signs of illness. Treatment controls clinical signs of illness, but does not eliminate infection. Treat mild cases by retrograde nasal flush with 1:10 enrofloxacin:water (2.27% injectable solution:tap water) q24h for at least 3 days after clinical signs have resolved. More severe cases usually respond well to systemic clarithromycin or azithromycin. Some clinicians report that danofloxacin by injection also resolved clinical signs.

Systemic amoebiasis

Various amoebae normally reside in the intestinal tracts of turtles and tortoises as commensals. Some can become opportunistic pathogens, and *Entamoeba* can be an important pathogen under certain host and environmental conditions. *Entamoeba invadens* is most pathogenic at 25°C (77°F), and is inhibited at 16°C (61°F) and 33°C (91°F). Disease usually occurs within a week or two after a period of hypothermia, when the environmental temperature of a reptile was not maintained in the optimal range. Clinical signs are nonspecific and include changes in daily activity patterns, spending more time than usual in the water dish, decreased appetite, and diarrhea. Death often occurs within a few days after the first subtle signs of illness occur. The infective stage of *Entamoeba invadens* is the quadrinucleate cyst, which is shed in the feces and can persist in the environment for extended periods. Ingested cysts develop into invasive trophozoites in the intestinal tract and can, like the cysts, can be seen under light microscopy. PCR is necessary to distinguish whether amoeba found on fecal examination should be treated or not. Typical gross postmortem signs of systemic amoebiasis include right sided to diffuse, grey-green, friable liver; fibrinous peritonitis, and petechial hemorrhages in the myocardium and kidneys. Cage mates of the deceased animal should be treated immediately with metronidazole (25 mg/kg PO q24h × 5 days) and broad-spectrum antibiotics because secondary bacterial infections are common. If appropriate for the species, the environmental temperature should be maintained at least 5-10°F above 25°C (77°F) during the treatment period. The environment of chelonians that have been diagnosed with a pathogenic amoeba infection should be thoroughly cleaned with hot (>65°C [150°F]), soapy water and disinfected with 1:5 household bleach:cool water solution for 20 minutes of contact time. Substrate, plants, cage furniture, and water dishes should be discarded.

Bacterial gastroenteritis/colitis

On occasion, bacterial colitis is diagnosed during the postmortem examination without identifying additional infectious agents. One reasonable explanation for bacterial infection of the gastrointestinal tract is an abnormally ineffective immune response, which allows normally commensal organisms to become pathogenic. Environmental conditions that may contribute to a weakened immune system include shipping, overcrowding, insufficient environmental resources (basking areas and shelters), and inadequate environmental complexity including tactile and visual components.

Non-infectious diseases

Non-infectious diseases at the Turtle Conservancy have included congestive heart failure in a Wolf Volcano Galapagos tortoise (*Chelonoidis nigra becki*), cloacal prolapse, beak overgrowth, and gout, both visceral and articular. Congestive heart failure, characterized by severe edema (anasarca), has been reported and one possible cause is an improper diet. The Wolf Volcano tortoise experiences extreme wet and dry seasons, so it is deprived of water under natural conditions for at least half of the year. Such water deprivation is rare in captivity, but could be an important physiologic need of this species.

Cloacal prolapse can occur in any species, and is most common in Burmese black mountain tortoises and radiated tortoises. In most cases the prolapsed cloaca can be reduced with granular sugar and steady digital pressure, and a purse-string suture around the vent is sufficient to prevent re-prolapse in the short term. A few cases have recurred more than one year after the first incident. Beak overgrowth, both rhinothecal and gnathothecal, is not rare in Burmese star tortosies, Burmese black mountain tortoises, and Chaco tortoises. Excessive beak can be trimmed in a manner similar to that employed with psittacine birds, using a stone bit on a high-speed rotary tool. Two of these tortoise species are known to eat a varied diet in the wild, and providing whole boiled eggs to the diet of Burmese black mountain tortoises and earthworms to the diet of Burmese star tortoises seems as if it might reduce or slow the reoccurrence of overgrowth. Gout is a somewhat enigmatic condition in reptiles. Young, growing Burmese star tortoises are particularly susceptible at the Turtle Conservancy, and visceral gout occasionally occurs together with cystic calculi.

Protozoal Parasitology for the Reptile Practitioner

Adolf K. Maas III, DVM, Dipl ABVP (Reptile and Amphibian), CertAqV

Session #220

Affiliation: ZooVet Consulting, PLLC, Bothell, WA 98041, USA.

When is a parasite not truly a parasite? Rather often in the field of herpetological medicine. The definition of a parasite is "an organism that lives in or on another organism (its host) and benefits by deriving nutrients at the host's expense." Presence of one organism on or within another organism does not define it, as the key portion of the definition is that the parasite creates an expense to the host with its presence. Cases where significant virulence or innate pathogenicity is identified should pursue treatment. How, then, should a patient be treated if there is no clear direct relationship of an expense to the host caused by the secondary organism in question?

This form of organism cohabitation is one of the most common health issues in reptile medicine and it is not common to find cases so simple that presence or cohabitation equals disease. All aspects of the health and condition of the patient must be considered including a review of the specific organism of concern as well as an attempt to define how the invader is affecting the host. In an effort to meet the standard of care, this needs to be performed before a need for treatment can be determined, let alone select specific therapies.

Evidence-based-medicine is critical to herpetological medicine, and to apply it in these situations the practitioner needs to form appropriate questions to determine if there is virulence involved with an infestation. These questions might include:

- Is the protozoan identified as a species known to have pathogenicity in the species of reptile involved?
- Is there physical damage to the host?
- Does the protozoan disrupt the host's energy and nutritional balance?
- Are there changes in the host's immunologic system?
- Is the host affected in other ways, in other systems (hematologic, renal, etc)?
- Is the presence of the protist associated with behavior abnormalities in the host?

These questions often have no well-defined answer, or at least not easily obtained answers. The identification of parasite species is often not answered, and there have been very few studies of protist-host relationships in reptiles. To further compound clear answers, in many cases the identified protist is also considered normal flora and the pathology seen is a consequence of high protozoal population densities rather than simple presence within the host.

Determination of a protozoan inhabitant as a pathogen or as a commensal is the first step to appropriate treatment, and this is anything but a well-defined process. As pathogenicity is specifically an innate function of the agent (the "pathogen") to cause damage to the host organism, it does not necessarily apply in reptile/protozoal relationships. Virulence is a more accurate term; it implies that the pathology can be variable depending on other factors, and that the invader does not have an innate pathology. Both commensalism and opportunism fit well under this definition. Thus, before a need for treatment can be assessed, the virulence of the identified protozoal agent needs to be determined, with a high virulence organism considered a parasite while a low virulence organism being commensal.

In reptilian host-protozoal relationships, the line between commensalism and parasitism often becomes rarefied. To have a true commensal relationship, there must be no detriment to the host organism, which is very uncommon and has been argued that it may not be possible. With alimentary protozoa, there may be situations where the organism is providing a benefit to the host (i.e. digestion of cellulose within the gut) and there is a net increase of positive effects to the host compared to the cost to the host by the inhabitation. However, the comparison between adverse effects and benefits is rarely direct, as increasing energy supply cannot be considered the opposite to the detriment caused by a vitamin deficiency. Thus, each situation must be examined independently as to the virulence displayed by the protist.

Superinfections appear to play a significant role in the virulence of protozoal flora in reptiles. In many species of reptiles, particularly herbivores and omnivores, protozoa live in limited, apathogenic relationships within the host's alimentary tract that benefits both the reptile as well as the protozoa. A superinfection would be defined by marked increases in protozoal population density. One method of the generation of superinfections is associated with repeated reinfections in less than a generation time frame. This occurs much more commonly in captive animals than in wild populations because of the nature of how these animals are housed. When animals are isolated in an enclosure in which they eat, drink, and defecate within a confined space, the result is that they ingest and repopulate their system with protozoa at higher densities than their wild counterparts would typically encounter.

Stress is another established etiology of superinfections, and it may be primary or in conjunction with reinfection rates. In any animal, changes in (captive) husbandry, poor environmental conditions, reproductive activity, seasonal cycles, underlying health conditions (infection, neoplasia, nutritional deficiencies), conspecific stress and competition, and many other factors all can cause immunosuppression and changes in metabolism. The corresponding physiologic changes can allow a commensal protist to replicate, cause pathology, and expand their tissue tropism, thus increasing their virulence. In short, when the populations of normal protozoa increase above a threshold, these organism's virulence damage the host, thus becoming parasites. Therefore, superinfections and the corresponding increase in virulence demonstrate the need for treatment despite not having an innately highly pathogenic organism as the root cause.

Determining a need for treatment must also involve evaluating the condition of the patient. It is exceptionally easy to identify a protist in a sample collected from a clinically healthy, normal patient, and decide that its presence equates a pathologic process of disease and thus make the decision for treatment. However, this approach must be questioned.

It is more difficult to examine a sample from a clinically ill reptile, identify the presence of protozoa, and determine whether or not the pathologic process is a result of said presence; however, as clinicians, we must be treating a patient, not a laboratory finding. This is the value of reviewing a complete anamnesis, having a good working knowledge of the natural history for that species, and selecting and properly executing appropriate diagnostic tools.

Although reptiles are well known for not showing overt symptoms in cases of disease, often the signs that the animal is presented for give us the clues that are needed to determine pathology. A thorough knowledge of a reptile's natural history, captive husbandry, behavior, and metabolism is necessary to effectively assess the health of these patients.

Ancillary testing can play a role in the determination of virulence and thus the need for treatment. Routine serum/plasma biochemistry testing is of limited value in specifically determining virulence, but can show changes in markers that indicate tissue damage. Furthermore, if the infestation is directly affecting specific organs, the biochemistry profile may reflect these effects.

Specific biochemical marker and function tests may be more indicative of damage to the host, and thus virulence. Serum electrophoresis has been demonstrated to provide more information on reptile immune system activity and can be an effective indicator of the effects of increased virulence. Urine gamma-glutamyl transferase (GGT) levels have been demonstrated to be a much more sensitive marker of early renal disease in birds and mammals and may be of similar value in reptiles. Serum and fecal biliverdin levels, while not commonly measured, may potentially be of higher diagnostic value in determining liver function than most other routinely run tests.

A complete blood count (CBC) can be very helpful in assessing virulence. Not uncommonly, a single CBC is performed on a clinically ill reptile patient and the results are reported as "within normal limits", causing many practitioners to put the value of hematologic evaluation low in their diagnostic plan. Granted, a single CBC performed at almost any time during the disease process has low diagnostic and prognostic value. However, the value of white blood cell assessments is in serial evaluations and monitoring trends and changes, rather than in a single static set of values. Additionally, specific cell type values that might be within reference ranges for a species can trend in directions that provide understanding of both pathologic processes as well as prognosis to the intuitive clinician.

Although non-invasive diagnostic techniques are preferential in most cases, adept reptile clinicians should pursue histopathologic diagnosis in complex, difficult and/or chronic cases. There are many excellent retrospective and review articles demonstrating virulence in protozoal infestations, and the results can be exceptionally useful both for planning treatment strategies as well as for understanding the prognosis of the patient.

Speciation of the purported pathogen may be necessary for effective management. If for no other reason, it may be necessary to determine if the protozoa found is an inhabitant or a pseudoparasite that is merely passing through the reptile's GI tract. Even true parasitism can be difficult to identify; *Sarcocystis* not only has sporozoites that can be difficult to identify, but an understanding of whether the reptile is playing the role of a primary or intermediate host, or if the organism is merely a pseudoparasite, needs to be assessed before appropriate treatment can be initiated. There are well-documented cases of reptile feces containing coccidian sporozoites that were parasites of their prey, but not known to inhabit the predator other than through GI transit.

Although most commonly associated with gastrointestinal infestation, protozoal parasites can inhabit and cause pathology in all tissues of a reptilian host. Treatment of protozoal parasites requires a careful identification of the parasite type, tissue tropism, associated pathology, and possible therapies. Therapeutic options must consider the overall health and condition of the animal as well as species sensitivities to particular medications, pharmacokinetics, and drug metabolism before appropriate therapy can be started. Every reptile clinician must take into account the entire host before selecting a chemotherapeutic agent.

Reptile Surgery and Endoscopy

Nicola Di Girolamo, DMV, MSc(EBHC), PhD, Dipl ECZM (Herpetology), and
Christoph Mans, Dr. med. vet., Dipl ACZM

Session #155

Affiliation: From the Clinica per Animali Esotici, Centro Veterinario Specialistico, Roma, Italy, and School of Veterinary Medicine, University of Wisconsin-Madison, 2015 Linden Drive, Madison, WI 53706, USA.

Abstract: Reptile surgery is a field in quick development. In recent years, new techniques have been developed for diagnosis and treatment of reptiles. In this masterclass we will illustrate current advances in surgical and endoscopic procedures in reptiles.

Skin Surgery

Reptile skin has the tendency to invert after incision, especially in squamates. Therefore, a slightly everting suture pattern (e.g., horizontal mattress) is recommended to ensure first intention wound healing. Healing of the skin can be accelerated if reptiles are maintained at the upper end of their preferred temperature range. Often, definitive skin healing with disappearance of the scab occurs after the first or the second ecdysis. Current recommendation for closure of coelomic breaches is to employ absorbable synthetic monofilament suture material (e.g., poliglecaprone 25, polyglyconate).

Surgery and Endoscopy of Chelonians

Aural abscesses

Aural abscesses are a common clinical presentation in aquatic turtles characterized by unilateral or bilateral swelling of the tympanic membranes. An association between aural abscesses and hypovitaminosis A has been proposed but could not be proven and the etiology of aural abscesses is more likely to be multifactorial. Surgical treatment consists of incision of the tympanic membrane and surgical debridement of the tympanic cavity under general anesthesia. A single vertical, two cross incisions, or a circular incision (180-360°) are made on the tympanum. By use of a cotton tip, debris and caseous material in the tympanum is gently removed. Culture and sensitivity testing may be performed from the debris or the tympanum. The tympanic cavity is lavaged, packed with an antimicrobial ointment and managed as an open wound to allow healing by second intention. Appropriate changes in management and diet are crucial to avoid recurrences.

Esophagostomy tube placement

Esophagostomy tube placement is typically recommended in anorectic chelonians, given the difficulty of oral administration of food and/or medication. Curved mosquito hemostats are inserted in the mouth and pushed on the side of the neck. Gentle pressure favors displacement of the jugular vein and carotid artery. A small skin incision is made over the tip of the mosquito exposing the external muscular layer of the esophagus. The esophageal wall is incised and the tip of the forceps is exposed. The feeding tube is grasped and passed through the incision and directed cranially. Once visualized through the mouth, the tube is gently curved and pushed into the esophagus towards the stomach up to the level of the predetermined length of the tube. A standard roman sandal suture (syn. Chinese finger knot) with an optional purse-string pattern around the tube is performed. Confirmation of the proper positioning of the tube is usually not necessary, but may be performed by administration of contrast media through the tube.

Celiotomy

In chelonians, the two main approaches to the coelom are plastron osteotomy and prefemoral fossa coeliotomy. With the increased availability of laparoscopic and endosurgical equipment, prefemoral fossa coeliotomy has gained popularity due to the reduced invasiveness compared to plastron osteotomy. However, in certain species and for certain surgeries, plastron osteotomy is still required as it permits a better maneuvering of cranial coelomic organs.

Prefemoral fossa coeliotomy

Prefemoral fossa celiotomy (syn. prefemoral coeliotomy) is currently considered the surgical access of choice for most reproductive surgeries and for diagnostic endoscopy. In particular, in semiaquatic and aquatic turtles, this approach avoids prolonged post-surgical dry-docking, which is necessary following plastron osteotomy. It should also be considered for urinary bladder and intestinal surgery. This approach is particularly indicated in species with a relatively small plastron and in semiaquatic and aquatic species. The chelonian is placed in ventral, dorsal or lateral recumbency depending on species, size, and indications for surgery. The skin, subcutaneous tissue, and the transverse and oblique abdominal muscles, and coelomic membrane are transected and the coelomic cavity accessed. The use of ring retractors is extremely useful to enhance access and visibility. Care should be taken to avoid trauma to the urinary bladder upon entering the coelom. Cystocentesis should be performed if the bladder is distended and impairs surgical access to the coelom. Closure of the body wall should be performed in 2-3 layers. It may not always be possible to close the coelomic membrane. Closure of the muscles and skin are routine. If a substantial amount of subcutaneous tissue is present, then additional subcutaneous sutures should be considered. The healing times after prefemoral coeliotomy (approximately 4 weeks) are substantially shorter than after a plastron osteotomy (approximately 1 to 2 years). In aquatic and semi-aquatic chelonia, early return to the aquatic environment is critical to allow normal behaviour, food intake and defecation. The prefemoral approach significantly shortens the post-surgical duration, during which time the turtle must be maintained out of water. Therefore, the prefemoral approach should be considered the preferred surgical approach to the coelom in aquatic and semi-aquatic chelonian species.

Plastron osteotomy

The size of the plastron osteotomy (syn. plastrotomy) is dependent on the indication for surgery (e.g., size of eggs, bladder stones, GI foreign bodies, etc.) and is limited cranially by the heart and caudally by the pelvic girdle. The animal is placed in dorsal recumbency. Various instruments (e.g., a rotary tool equipped with a cutting circular blade, an oscillating sagittal saw, etc.) may be employed to create the incision into the plastron. Three sides of the flap should be incised at a 45-degree angle to obtain slightly beveled incisions. The fourth side is only partially incised and is used as a hinge. A periosteal elevator is used to complete the three full-thickness incisions of the plastron and to elevate the flap. The flap is reflected cranially or caudally and covered with moist gauze. The coelomic membrane is visualized and a ventral midline incision is performed taking care to avoid the ventral abdominal veins. After plastron osteotomy, the coelomic membrane is gently sutured with a fine monofilament absorbable suture in a simple interrupted or continuous pattern. The lack of suturing of the coelomic membrane has been suggested to be associated with increased risk of postoperative adhesions. The bony flap is repositioned and sutured (in young or demineralized chelonians) or stabilized by means of epoxy resins, fiberglass mesh, or metal plates and screws. Plastron osteotomy is generally associated with prolonged surgical procedure time, prolonged recovery and is thought to be significantly more painful compared to the prefemoral soft tissue approach. The plastron flap may become a sequestrum and provide temporary protection to the developing bone. Eventual post-surgical complications are usually serious, require prolonged treatment and include: lack of revascularization of the bone flap and consequent necrosis, infection with consequent coelomitis, and dehiscence of the bone flap margins. A further limitation of the technique is the need for dry-docking of aquatic turtles. Even with application of a fiberglass patch postoperative leakage may occur, leading to infection. Healing times of the bone flap are variable, but are considered to be around 1 to 2 years.

Endoscopy-assisted prefemoral oophorectomy and salpingectomy

A technique for exteriorization and excision of ovaries and oviducts of chelonians through the prefemoral fossa has been described. The technique relies on the assistance of standard endoscopic equipment (e.g., 2.7-mm, rigid endoscope). Depending on species, size and individual morphology, oophorectomy, salpingectomy, and/or salpingotomy may be performed through the prefemoral approach. Chelonians are preferably placed in dorsal recumbency, although ventral, or lateral recumbency may be indicated in particular instances (e.g., unilateral egg dystocia). A rigid endoscope is introduced into the coelom following a standard prefemoral coeliotomy approach. The reproductive tract is identified and gently grasped with atraumatic grasping forceps preferably by an avascular connective area of the ovary. Care must be taken to avoid rupture of ovarian follicles. Once all ovarian follicles are exteriorized and the mesovarium is visible, the ovarian vasculature in the mesovarium is ligated and transected. Hemostasis and complete excision of the ovarian tissue should be confirmed with the endoscope. If the oviduct is diseased and surgical removal is intended, the ipsilateral oviduct is exteriorized, ligated and transected through the same prefemoral incision. The procedure is then repeated for the contralateral ovary and oviduct. Often the procedure may be performed for the contralateral ovary via the same prefemoral incision. However, performing salpingectomy or salpingotomy of the contralateral oviduct is usually not possible, and coelomic access is required through the contralateral prefemoral fossa. Retained eggs in the oviduct and ectopic eggs free within the coelom can also be removed using this technique.

Cystotomy

In general, to resolve the presence of cystic calculi or ectopic eggs in the urinary bladder in chelonians, the surgical procedure should be elected in the following order: (1) Transurethral endoscopic retrieval, mechanical destruction (eggs) or lithotripsy (calculi); (2) Pre-femoral fossa coeliotomy and cystotomy; (3) Standard coeliotomy and cystotomy. Removal of uroliths through a prefemoral fossa approach is not always feasible. In a case series of 10 desert tortoises (*Gopherus agassizii*) good candidates for the prefemoral fossa approach were large tortoises (over 15 cm in carapace length), with small calculi (less than twice the length of the fossa) that did not have a laminated radiographic appearance. Note that in the pre-operative evaluation, radiographs, computed tomography, ultrasound and cystoscopy are essential to reach a definitive diagnosis. In rare instances, (e.g., neoplasms, chronic prolapses of the urinary bladder) partial cystectomy may be indicated.

Prefemoral fossa orchiectomy

Prefemoral endoscopic orchiectomy has been recently described in turtles and tortoises. Briefly, the testis is visualized and grasped, distending the mesorchium. Ligation clips or radiosurgical instruments are employed for hemostasis and dissection of the ligament. The testis is removed while the epididymis is left in situ. Orchiectomy may be also performed through the prefemoral fossa approach without endoscopic guidance. An incision is performed in the prefemoral fossa and the musculature is dissected. Urinary bladder, intestine and lungs are manipulated in order to expose the testes. The testis is grasped using a curved hemostat, the spermatic cord is dissected and ligation clips are placed around the spermatic cord and testicular artery. The spermatic cord and the testicular artery are dissected distal to the ligation clips. The same technique may be performed under endoscopic assistance with exteriorization of the testes. The testis is visualized through prefemoral fossa coelioscopy.

The testis is grasped and exteriorized. Ligation clips or radiosurgery are employed to excise the mesorchium preventing bleeding. The epididymis is controlled with the endoscope for hemorrhages. The coelomic access is routinely closed.

Cystoscopy in chelonians

Historically, cloacoscopy and cystoscopy have been used for evaluation of intrinsic disorders of the cloaca, urinary bladder, and accessory bladders. More recently, a novel diagnostic application of cystoscopy has been described utilizing the transparency of the organ to visualize the coelom. The value of cystoscopy as a diagnostic tool relies on two main features of the chelonian urinary bladder, the large size and its transparency. Chelonians may be positioned in dorsal or ventral recumbency for cloacoscopy. Chelonian cystoscopy is indicated for chelonians with a non-specific illness, as an initial screening technique to identify the diseased organ. It can be used in chelonians with dystocia, in order to screen for ectopic eggs in the bladder and chelonians with uroliths, for diagnosis and treatment, and in chelonians that underwent trauma, to evaluate bladder integrity, as well as in selected species of chelonians lacking sexual dimorphism or for selected immature chelonians, in order to identify the sex.

Technique: A rigid or flexible endoscope is gently inserted in the vent and directed cranially in order to reach the cloaca. The cloaca is a sac-like cavity into which the ureters, gonadal ducts, colon, and bladder(s) empty. In some aquatic turtles, paired sac-like structures, the accessory bladders, open into the cloaca dorsal to the colon. The endoscope, once in the cloaca, is directed toward the urethral opening, which is located ventral to the rectum. Access to the urethral sphincter is gained by gentle ventral pressure. Warm (30°C) fluids are infused (1 drop every 3-4 seconds) during the entire procedure and to allow distension of the urethral opening. As an alternative, air can be insufflated. Care should be paid to avoid over-distension of the bladder in smaller chelonians. Gentle pressure and fluid infusion are usually sufficient to gain access to the urinary bladder.

Coelioscopy in chelonians

A standard left-sided approach is performed positioning the animal in right lateral recumbency and holding it by means of a vacuum positioning aid or sand bags. The left hindlimb is retracted and taped caudally in order to expose the left prefemoral fossa. Surgical scrub is performed on the prefemoral fossa and the chelonian is draped. A 3- to 4-mm skin incision is performed in a cranio-caudal direction in the skin in the center of the prefemoral fossa. The subcutaneous fat and connective tissues are bluntly dissected and the coelomic membrane is gently stabbed with a hemostat. The endoscope is inserted at a 30-45° angle and directed cranially to visualize the liver, gallbladder, stomach, pancreas, duodenum, heart. The testes (or ovaries), epididymides and adrenals are visualized by dorso-caudal direction of the endoscope. The kidneys lay in the retroceolom and are visualized after dissection with an endoscopic forceps or with laser of the coelomic membrane, right after the epididymis in male chelonians. Diagnostic samples can be obtained by use of an endoscopic biopsy forceps. After inspection of all the organs of interest, the coelom is gently sutured with an interrupted pattern and the prefemoral skin is routinely sutured.

Surgery of Squamates

Phallectomy

Male snakes and lizards have paired copulatory organs (i.e., hemipenes) that lie in respective sacs caudal to the cloaca in the ventral tail. Phallectomy is employed to resolve hemipenile disorders. Hemipenile prolapse is a very common presentation. If the tissue is already necrotic or the prolapse is recurrent, amputation is required. Amputation is performed *in toto*, as hemipenes do not contain the urethra. In squamates, amputation of a single hemipene does not preclude reproduction, while in chelonians phallectomy precludes reproduction. The reptile is positioned in dorsal recumbency and the hemipene is surgically prepped. In small individuals, two transfixing ligatures are placed at the base of the hemipene. The tissue is excised distal to the ligatures. The hemipenile stump can be closed with a simple continuous or a purse-string suture, in particular in larger species.

Limb amputation

Indications for amputation of limbs in lizards and chelonians are severe trauma or infection, severe joint infections, which are refractory to treatment, as well as non-healing chronic fractures or neoplasia. In general, chronic infected distal limb wounds and joint infections, in particular with associated osteomyelitis usually do not respond to medical therapy and wound management, and instead limb amputation should be considered. Amputation of the forelimbs should be performed through the scapulohumeral joint and amputations of the hindlimbs through the coxofemoral joint. Midshaft amputation of the humerus or femur are not recommended, since it is very likely that the remaining limb stump will be traumatized by attempted ambulation. In female reptiles used for breeding purposes or planned to be released in the wild, amputation of the hindlimb might interfere with successful reproduction, since digging of nesting sites may be impaired. Limb amputation is performed using the same techniques as in mammals. Complications are uncommon.

Tail amputation

Amputation of an infected, traumatized or necrotic tail is a common problem in lizards. Surgical amputation of the tail is recommended in order to achieve primary wound healing following surgical resection of the diseased distal tail and closure of the amputation site. Radiographs should be taken prior to surgical amputation in order to evaluate for underlying bone involvement. Amputation of a significant portion of the tail in arboreal lizards (e.g., green iguanas, *Iguana iguana*), has substantial effects on their ability to balance and climb. Therefore, owners should be informed that the enclosure may require adjustment following tail amputation and that changes in locomotion are possible. The patient is anesthetized and placed in ventral recumbence. The tail amputation site is surgically prepped and a tourniquet is placed. The location of the hemipenes should be considered in cases of proximal tail amputation. Symmetric wedge incisions of the skin are made on lateral aspects of the tail in lizards with laterally flattened tails (e.g., iguanas) or on the dorsal and ventral aspect in lizards with dorso-ventrally flattened tails (e.g., bearded dragons, *Pogona* spp.). The skin incision should be made distal enough to allow for tension free wound closure following amputation, but at the same time be proximal enough to avoid incomplete excision of diseased tissue. The soft tissue and bone are transected. The ventral tail vein may require ligation in larger lizards. The amputation site is lavaged and assessed for hemorrhage following release of the tourniquet. The muscles should be apposed over the vertebrae with simple interrupted sutures. The skin edges are trimmed and closed with horizontal mattress sutures. The amputation site can be bandaged for the first few days following surgery, in order to aid in absorption of wound discharge. Suture removal is recommended 4-6 weeks following amputation. In species that performs tail autotomy, tail amputation can be performed exploiting the natural line of fractures on the vertebrae by use of quick manual latero-lateral movements. Regrowth of the tail is possible after surgical amputation, but not commonly seen.

Coeliotomy

Surgical approaches to the coelom greatly vary depending on the species. Appropriate preparation of the reptile patient for coeliotomy is mandatory. Sterile scrub brushes may be employed to provide effective cleaning of reptiles. Once anesthetized, the patient is instrumented, placed in the recumbency indicated for the surgical procedure and surgically prepped. In general, the size of the incision will depend on the indications for coeliotomy. During the opening of the coelomic membrane, the surgeon should watch for signs of free gas or liquid in the coelom, associated with gastrointestinal (GI) tract and urinary bladder perforation, respectively. After entering the coelom, all of the organs that are visible should be carefully inspected.

Lizards: In lizards, there are three main approaches to the coelom including paramedian, median and flank approaches. The paramedian and median approaches are generally indicated in lizards that are dorsally compressed (e.g., families *Iguanidae, Agamidae, Gekkota*). In chameleons and other lizards that are laterally compressed (e.g., basilisks), the coelom may be easily approached through the flank. The paramedian approach is generally preferred over the median approach to avoid the ventral midline abdominal vein, a large vessel that runs just over the linea alba. An incision of the skin is made parallel to the midline. The incision may be made with a scalpel blade or dissecting devices (e.g., lasers, radiosurgical and electrosurgical equipment). In a study conducted on green iguanas, radiosurgery and laser both produced bloodless incisions, but radiosurgery caused significantly less collateral tissue damage in the skin and the muscle. The distance of the incision from the midline will depend on the size of the lizard. The incision is extended cranially and caudally using scissors, taking care to avoid the ventral abdominal vein. In some instances, the ends of the incision may need to be extended at a 90º angle, forming a sort of "L" or "H". The skin is retracted and the musculature is dissected by use of blunt scissors, cotton-tipped applicators, laser or electrosurgical equipment. Sharp dissection of the musculature should be avoided in order to minimize bleeding.

Snakes: In snakes, a paraventral coeliotomy is indicated for most surgeries. Obviously, a single incision does not permit access to all the organs. Therefore, an incision needs to be made at the level of the organ of interest. The skin is incised between the first and second row of lateral scales, and the resulting surgical wound is scalloped. This incision avoids distortion of the ventral scutes once the skin is sutured in an everting pattern.

Closure of the coelom in snakes and lizards: In squamates, the coelomic membrane and muscular wall are fragile and do not hold sutures. Therefore, closure of the coelom relies on the skin sutures. In small to medium sized squamates, the coelomic membrane, muscularis and subcutis are sutured together in a simple continuous suture pattern with monofilament absorbable sutures. The skin is sutured with fine monofilament nonabsorbable sutures (e.g., nylon) in an everting pattern as previously discussed.

Cystotomy

Cystotomy may be indicated for removal of cystic calculi and ectopic eggs from the urinary bladder. In recent years alternative techniques to routine cystotomy, in particular in chelonians, have been reported, which are less invasive. These less-invasive techniques include transurethral endoscopic techniques (cystoscopy) or prefemoral endoscopy-assisted cystotomy. However, if such approaches fail (e.g., when dealing with cystic calculi of large size) cystotomy may be required. Access to the coelom is routinely performed through a paramedian incision in lizard. In small-to medium species, manipulation and exteriorization of the urinary bladder should be reduced to the minimum, due to the risk of trauma of the thin bladder wall. Stay sutures on the urinary bladder wall and moist gauze around the surgical site minimize urine leakage into the coelom. In lizards, a longitudinal incision is made in the ventral aspect of the bladder. The bladder stones or eggs present in the urinary bladder are removed. Retrograde use of an endoscope may be useful to ascertain the removal of all the material. The urinary bladder wall is closed with a monofilament suture in a single or double inverting layer. A rounded, atraumatic needle and minimal traction are mandatory to avoid rupture of the urinary bladder wall during suture in small-to-medium sized reptiles.

Nephrectomy

In snakes, neoplasms or degenerative disorders of the kidneys are not uncommon, and unilateral nephrectomy is associated with an acceptable prognosis if the remaining kidney is functional. Often, a swelling of the caudal third of the snake is present. Standard coeliotomy is performed at the level of the affected kidney. The contralateral kidney should be grossly inspected. The affected kidney is isolated from the surrounding tissue. The renal vein and the renal arteries are ligated. The ureter is ligated distally and the kidney removed. A unilateral gonadectomy may need to be performed as the vas deferens in males and the oviduct in females are in contiguity with the kidney.

References

1. Bronson E, Pereira M, Sanchez C, et al. Iridophoroma in a veiled chameleon, *Chameleo calyptratus*. *J Herp Med Surg*. 2006;16:58-60.

2. Di Girolamo N, Mans C. Reptile soft tissue surgery. *Vet Clin North Am Exot Anim Pract*. 2016;19:97-131.

3. Funk RS. Early neutering of male green iguanas, Iguana iguana—an experiment. *Proc Annu Conf Assoc Rept Amph Vet*. 2000;147-149.

4. Govett PD, Harms CA, Linder KE, et al. Effect of four different suture materials on the surgical wound healing of loggerhead sea turtles, *Caretta caretta*. *J Herp Med Surg*. 2004;14:6-11.

5. Hannon DE, Garner MM, Reavill DR. Squamous cell carcinomas in inland bearded dragons (*Pogona vitticeps*). *J Herp Med Surg*. 2011;21:101-106.

6. Innis CJ, Feinsod R, Hanlon J, et al. Coelioscopic orchiectomy can be effectively and safely accomplished in chelonians. *Vet Rec*. 2013;172:526.

7. Innis CJ, Hernandez-Divers S, Martinez-Jimenez D. Coelioscopic-assisted prefemoral oophorectomy in chelonians. *J Am Vet Med Assoc*. 2007;230:1049-1052.

8. Mans C, Sladky KK. Diagnosis and management of oviductal disease in three red-eared slider turtles (*Trachemys scripta elegans*). *J Small Anim Pract*. 2012;53:234-239.

9. Mans C, Sladky KK. Endoscopically guided removal of cloacal calculi in three African spurred tortoises (*Geochelone sulcata*). *J Am Vet Med Assoc*. 2012;240:869-875.

10. McFadden MS, Bennett RA, Kinsel MJ, et al. Evaluation of the histologic reactions to commonly used suture materials in the skin and musculature of ball pythons (*Python regius*). *Am J Vet Res*. 2011;72: 1397-406.

11. McFadden MS. Suture materials and suture selection for use in exotic pet surgical procedures. *J Exot Pet Med*. 2011;20:173-181.

12. Myers DA, Wellehan JF Jr, Isaza R. Saccular lung cannulation in a ball python (*Python regius*) to treat a tracheal obstruction. *J Zoo Wildl Med*. 2009;40:214-216.

13. Selleri P, Di Girolamo N, Melidone R. Cystoscopic sex identification of posthatchling chelonians. *J Am Vet Med Assoc*. 2013;242:1744-1750.

14. Smith DA, Barker IK, Allen B. The effect of ambient temperature and type of wound on healing of cutaneous wounds in the common garter snake (*Thamnophis sirtalis*). *Can J Vet Res*. 1988;52:120-128.

Getting to Grips with the Reptile Environment

Stephen J. Divers, BVetMed, DZooMed, Dipl ACZM, Dipl ECZM (Herpetology), Dipl ECZM (Zoo Health Management), FRCVS, and
Scott Stahl, DVM, Dipl ABVP (Avian)

Session #184

Affiliation: From the Department of Small Animal Medicine and Surgery (Zoological Medicine), College of Veterinary Medicine, University of Georgia, 2200 College Station Road, Athens, GA 30602, USA.

Abstract: The majority of reptile diseases are either directly caused or predisposed by poor captive management. Many aspects of captive husbandry have a direct effect on the health and welfare of reptiles. Unfortunately, welfare legislation for this group of vertebrates in the U.S. lags behind many other nations. This roundtable provides an opportunity for the presentation and discussion of various issues facing the provision of appropriate environments for reptile patients.

Introduction

The majority of reptile disease presentations are either directly or indirectly caused or exacerbated by poor husbandry and nutrition.[1-3] In addition, poor environment has both immediate and delayed effects on health and welfare.[4,5] A veterinarian has little chance of successfully treating reptiles if he/she does not appreciate the importance of the environment, and the modern approach to providing such.[6]

Species

The importance of identifying the species (and subspecies) cannot be overemphasized. In certain cases, even locality may cause diet and environmental preferences within the same species. Different species from different geographic locations must never be mixed. Ideally, only a single species should be kept in any enclosure, and care must be exercised to avoid competition for resources such as food, basking areas and retreats. Some snakes may eat cagemates (e.g., kingsnakes), many male lizards are territorial and aggressive (e.g., leopard geckos, bearded dragons), while some species (*Chameleo* spp) are so territorial that individual isolation is often essential.

The Vivarium Enclosure

The type of enclosure (arboreal, terrestrial, subterranean and aquatic) should be appropriate for the species. Arboreal species (e.g., green iguana, *Iguana iguana*; emerald tree boa, *Corallus caninus*) prefer to move in a vertical plane while terrestrial species (e.g., skinks, *Tiliqua* spp; rat snakes, *Elaphe* spp) prefer a more down to earth existence. The size of the enclosure is also important and although many breeders and retailers may be able to intensively manage stock the author prefers to advise clients on minimum enclosure sizes and the importance of providing the largest enclosure possible. There are very few guidelines available for minimum space requirements.

Vivarium construction

Glass aquaria are commonly used but the greater visualization perceived as an advantage to the owner may serve as a stressful reminder of open exposure and poor security to the inmate. Glass is a poor insulator and greater heat loss may lead to dramatic temperature fluctuations. Even if the entire top of the enclosure is covered by mesh, ventilation may be severely reduced due to a lack of ventilation along the sides and back wall. Plastic-coated wooden, plastic or fiberglass enclosures are popular because they are more versatile, permitting the incorporation of additional ventilation panels, front access via sliding glass doors and greater security for the inhabitants.

Cage furnishings

In most cases what is placed inside the vivarium is for the aesthetic enjoyment of the owner rather than meeting the biologic and ethologic demands of the reptile. Particulate floor substrate is often changed infrequently due to their relative high cost. Some materials may cause contact dermatitis, acute stomatitis or intestinal impaction is ingested. The provision of hideouts or retreats are often overlooked and when provided may be too spacious to provide adequate security. Newspaper, Astroturf and commercial organic particulates (e.g., bark chips) are suitable materials, but they must be replaced regularly. Soil, sand, natural leaf litter can also be used but care is required to maintain hygiene and not to introduce disease. Sterilization of such material can be easily accomplished by baking in an oven. Gravel and pebbles are not recommended as they are difficult to clean and often ingested. Other essential items include a water bowl (large enough for reptile to bathe), and secure branches (non-toxic fruit trees) for arboreal species.

Vivarium hygiene

It is an unfortunate fact that many owners will decide that it is time to clean the vivarium once a foul odor becomes obvious, and therefore poor hygiene. Some reptiles will naturally excrete, or can be trained to excrete in one area. Effective hygiene requires the removal of organic debris prior to disinfection. All chemicals, even if considered safe, should be thoroughly rinsed away before replacing the reptile(s).

The Captive Environment

Captive environments may be considered as tropical, temperate, desert, rainforest, scrubland etc. These ecosystems can be notoriously difficult to reproduce in captivity. In certain areas the local weather may permit the keeping of reptiles in outdoor enclosures for at least some of the year. Theft, predators and wildlife carriers of disease should be considered.

Heating

There are a variety of heaters that may be employed including incandescent bulbs, infrared ceramics, heating pads or mats, warming cables, tubular heaters, radiators, convector heaters and natural sunlight radiation. From a clinical perspective there are several factors that must be considered. Is the heater screened from the animal to prevent burns? Is the heating equipment sufficient to provide the preferred optimum temperature zone (POTZ) for the reptile? Is the heating provision suitable to the species in question? Most arboreal species thermoregulate in a vertical plane and therefore heating should be positioned in the roof of the enclosure. Terrestrial species will tend to move in a more horizontal plane and so all heating should be concentrated at one end and not evenly distributed throughout the length of the vivarium. Finally, captive aquatic environments seldom permit the establishment of a water temperature gradient but acceptable water temperatures should be provided using a dedicated double-insulated water heater. Electrical cords must be protected from the powerful bites of aquatic chelonians as electrocution is not only fatal to the kept but presents serious threats to the keeper as well.

Temperature control: High wattage heaters must be controlled as at times only a fraction of their output may be required to maintain the desired thermal gradient. Many thermostats have the ability to carefully regulate temperature using a dimming system of gradual wattage reduction. Thermostats with nighttime drop facilities are useful to create a natural diurnal temperature fluctuation. Thermostat failure can be disastrous, especially if the power to the heater(s) is maintained after failure.

Temperature gradient: Although the temperature readings on modern-day thermostats are generally trustworthy, it is still advisable to measure the temperature gradient within the enclosure using a dual read-out digital thermometer, preferably with memories for maximum and minimum temperatures. A single temperature is meaningless in most cases; the thermal gradient within the vivarium is what is important. The daytime basking temperature (hottest daytime temperature), daytime air temperature gradient and nighttime air temperature gradient should be elucidated during the consultation; however, be aware that many owners will reel off the known requirements for their particular species without actually having measured the temperature with an accurate thermostat. It is important not to lead the client when asking husbandry questions, nor to force an inaccurate answer.

Environmental Lighting

Most snakes do not have any significant requirement for broad or full-spectrum lighting; however, many diurnal lizards and chelonia do require quality light and in particular wavelengths of around 295 nm (UVB) for the production of vitamin D_3 that plays an essential role in calcium metabolism. There is also some indication that natural light has behavioral and psychologic benefits. By far the best source of quality full-spectrum lighting is unfiltered sunlight; however, there is an increasing number of fluorescent and even LED lights commercially available but the spectral quality varies dramatically between different makes. More worrisome is that several lights specifically marketed as full-spectrum lights for reptiles are anything but, and this can create confusion for both veterinarians and clients alike. In general, for basking species the use of mercury halide lamps can provide light and heat up to a distance of 1 meter. Fluorescent strip-lights, compact fluorescents, and LEDs that specifically indicate a significant UVB output can provide light (without significant heat) over shorter distances of up to 30 cm.

Most transparent plastic and glass barriers filter out UVB wavelengths and therefore placing a reptile in a greenhouse or conservatory is not acceptable unless specialized UVB-transmitting materials are used. In order to determine whether the lighting is suitable for a diurnal reptile, the owner must be quizzed as to the exact photoperiod, age, position, make and model of light being used within the vivarium. In many cases, captive lighting is deficient in one or more of these facets.

Humidity

This is often an overlooked aspect of the captive environment with even the most enthusiastic pet reptile keeper failing to install a hydrometer within the vivarium. The importance of humidity is also poorly appreciated, for example, the high incidence of renal disease in adult iguanas may, in part, be related to chronic dehydration from prolonged inadequate humidity. Humidity is seldom directly controlled, although the advent of dedicated humidifiers and sprinkler systems is changing this perception. Instead humidity tends to be a function of temperature, water surface area and ventilation. The major problem in most cases is the view that ventilation must be decreased to increase humidity. Such actions result in stagnant air and an increase in bacterial and fungal infections, especially of the integument and respiratory system. Far better to maintain ventilation but maintain humidity by providing a greater water surface area (shallow water containers, water drip systems, sprinkler systems, regular spraying) or an increase in water temperature by placing a heat mat under a shallow water container.

References

1. Divers SJ. Reptiles: Husbandry, history taking and clinical examination. *In Pract.* 1996;18:51-65.

2. Mader DR. *Reptile Medicine and Surgery.* Second ed. Philadelphia, PA: WB Saunders Co;2006.

3. Raiti P, Girling S. *Manual of Reptiles.* 2nd ed ed. Cheltenham, England: British Small Animal Veterinary Association;2004.

4. Warwick C, Arena P, Lindley S, Jessop M, Steedman C. Assessing reptile welfare using behavioral criteria. *In Pract.* 2016;35:123-131.

5. Warwick C, Frye FL, Murphy JB. *Health and Welfare of Captive Reptiles.* London:Chapman & Hall;2004.

6. Barten SL, Fleming GJ. Current herpetological husbandry and products. In: Mader DR, Divers SJ, eds. *Current Therapy in Reptile Medicine and Surgery*. St Louis, MO: Elsevier;2014:2-12.

Overview of Reptile Clinical Pathology

Stephen J. Divers, BVetMed, DZooMed, Dipl ACZM, Dipl ECZM (Herpetology), Dipl ECZM (Zoo Health Management), FRCVS, and
Scott Stahl, DVM, Dipl ABVP (Avian)

Session #185

Affiliation: From the Department of Small Animal Medicine and Surgery (Zoological Medicine), College of Veterinary Medicine, University of Georgia, 2200 College Station Road, Athens, GA 30602, USA.

Abstract: Reptile clinical pathology is as much an art form as it is a science! Indeed, when you start correlating your pathology and histopathology findings to hematology and plasma chemistry changes you might be tempted to throw in the towel. This roundtable focuses on the appropriate collection, processing, and interpretation of samples, and specifically tries to bring some of the recent research to light.

Introduction

The accumulation of species-specific reference ranges continues to be slow for most reptiles. Unfortunately, most publications fall short of the standardized requirements necessary to establish true reference ranges, and some fail to take into account the significant effects of gender, age, season, reproductive status, diet and feeding schedules.[1] As zoological practitioners we must adhere to the established procedures in domestic animal clinical pathology, particularly regarding consistent collection, handling and laboratory practices, and deviate only when proven necessary to do so.

Despite the recent interest and reviews in reptile clinical pathology, much detailed data can still only be found outside of the veterinary literature.[2,3] Given the frequency with which the clinician may be forced to evaluate a clinical pathology report without an appropriate reference range, referring to some of these older but detailed chapters can still be rewarding.[4-6] Due to necessary space constraints, this chapter is an abbreviated summary, and reference to more detailed clinical pathology texts is suggested.[2,3,7,8]

Sample Collection

The blood volume of reptiles varies between 4 and 8% of bodyweight, and it is generally accepted that 10% of total blood volume (i.e., 0.4-0.8 ml/kg) can be safely collected from healthy reptiles.[9] Lymphatic contamination is a distinct possibility during reptile venipuncture. Complete or selective hematologic and biochemical evaluations are possible but the accuracy of results is often adversely affected by poor venipuncture, handling or laboratory skills. The production of poor quality blood data can be useless or even worse, misleading to the clinician. Venipuncture is generally a blind technique in reptiles and as a general guide up to 0.4-0.5 ml/100 g may be safely collected from any healthy reptile – less in debilitated animals. The jugular vein of certain tortoises may be visible particularly following temporary caudal occlusion of the vessel but in any case anatomic knowledge of the position of veins is vital. It is wise to aseptically prepare the venipuncture site, as infection and abscessation following venous access is a potential complication. Given the variability in published ranges, more reliance should be placed upon establishing an individual's observed range and using serial sampling to monitor the progress of hematologic and biochemical changes rather than relying on a single result. The following descriptions indicate the most humane and practical venipuncture sites.

Snakes

The two common sites for venipuncture in snakes are the caudal (ventral tail) vein and the heart. The caudal vein is accessed caudal to the cloaca, between 25 and 50% down the tail. It is wise to avoid the paired hemipenes of males (that may extend up to 14-16 subcaudal scales down the tail), and the paired cloacal musk glands of females (that may extend up to 6 subcaudal scales). The needle is angled at 45-60° and positioned in the ventral mid line. The needle is advanced in a craniodorsal direction, while maintaining slight negative pressure. If the needle touches a vertebral body it is withdrawn slightly and redirected more cranially or caudally. This vessel is most easily entered in larger snakes and lymphatic contamination is possible. With the snake restrained in dorsal recumbency, the heart is located approximately 22-33% from snout to vent. Sedation is preferred. The heart is palpated and immobilized. The needle is advanced at 45° in a craniodorsal direction into the apex of the beating ventricle. Blood often enters the syringe with each heartbeat. It is wise to maintain digital pressure for 30-60 seconds following this technique. This technique appears to be safe in snakes of all sizes and is less likely to result in lymphatic contamination. It has been employed in snakes of all sizes from 10-g neonates to 100-kg constrictors. Nevertheless, good restraint is essential if significant cardiac trauma is to be avoided.

Lizards

The most clinical useful vessel is the caudal (ventral tail) vein. The needle is positioned in the ventral midline between 20%-80% down the tail and advanced at 45-90° in a craniodorsal direction, while maintaining slight negative pressure. If a coccygeal vertebra is encountered the needle is withdrawn slightly and redirected either further craniad or caudad. An alternative approach that is particularly useful for the larger or more ventrodorsally compressed species, is to insert the needle from the lateral midline. The needle is advanced at 45-90° in a craniomedial direction, aiming just ventral to the lateral processes of the coccygeal vertebrae. Lymphatic contamination is more likely with the lateral approach. Lizards possess a large ventral abdominal vein that runs within a suspensory ligament just below the linea alba. This vessel is most easily entered in the mid-coelomic region. The needle is positioned in the ventral midline and advanced in a craniodorsal direction. It is difficult to apply pressure to this vessel, which makes post-venepuncture hemorrhage a potential complication.

Tortoises, turtles and terrapins

The most clinically useful vessels are the jugular, subcarapacial, and dorsal coccygeal veins. The left and right jugular veins are preferred because of the greatly reduced risk of lymphatic contamination. The regional anatomy varies with species but the vessel is generally located laterally and may even be visible if temporarily occluded by digital pressure at the base of the neck. The needle is positioned caudal to the tympanum, and directed in a caudal direction. Post-venepuncture pressure should be applied to prevent hematoma formation.

A subcarapacial site is also available and formed by the venous communication between the most craniad intercostal vessels arising from the paired azygous veins and the caudal cervical anastomosis of the left and right jugular veins. This sinus can be accessed with the chelonian's head either extended or retracted, making it useful for uncooperative or aggressive individuals. Depending upon the species and conformation of the carapace, the needle may be bent up to 60° and positioned in the mid-line just caudal to the skin insertion on the ventral aspect of the cranial rim of the carapace. The needle is advanced in a caudodorsal direction maintaining slight negative pressure. If a vertebra is encountered the needle is withdrawn slightly and redirected further craniad or caudad. Lymph contamination appears to be a common complication. The dorsal coccygeal vein is probably the most commonly employed of the tail veins. The needle is angled at 45-90° and placed, as craniad as possible, in the dorsal mid line of the tail. The needle is advanced in a cranioventral direction while maintaining slight negative pressure. If the needle encounters a vertebra it is withdrawn slightly and redirected more craniad or caudad. The exact position, size and even presence of this vessel may vary between species and there is a significant risk of lymphatic contamination.

Crocodilians

The most appropriate venepuncture sites are the caudal (ventral tail) vein in small to medium crocodilians and supravertebral vein in medium to large specimens. The supravertebral vein is approached with the needle positioned in the dorsal midline, just caudal to the occiput and perpendicular to the surface of the skin. The needle is slowly advanced while maintaining slight negative pressure. Excessive penetration may lead to spinal trauma. The technique for caudal vein access is as described for lizards. The heart is located on the ventral midline, approximately 11 scale rows caudal to the forelimbs. Cardiocentesis in crocodilians is not as safe as it is in snakes because the heart cannot be stabilized.

Sample Handling

Traditionally, the anticoagulant of choice has been heparin because of the perceived risks of lysis caused by EDTA; however, studies have demonstrated the superiority of EDTA in numerous species of lizards and snakes, and even some chelonians.[10,11] Therefore, species-specific anticoagulant preferences should be considered. A fresh blood smear should be made at the time of venipuncture using blood that has not been contaminated by anticoagulant (i.e., the last drop in the syringe). Of the various techniques that have been employed to create blood smears coverslip-to-slide and bevel-edge slide technique have been found to create a preferred monolayer with reduced numbers of smudge cells.[12] Blood for hematologic evaluation should ideally be processed within 1 hour. Delays can be accommodated by refrigerating the sample, but deterioration is inevitable. Blood smears should not be refrigerated as water condensation on the glass may damage cell morphology. Blood should never be frozen as this will result in cell lysis.[13] For biochemical analyte determinations, heparin is generally preferred over serum because (a) it results in a greater amount of sample per unit of blood, (b) can be processed immediately without waiting for 30-60 minutes or more for clot formation, and (c) prevents the formation of fibrin clots that can interfere with many automated analyzers. A recent study that evaluated the differences between serum and plasma in green iguanas demonstrated that statistically significant differences did exist between certain analytes.[13] Total protein, albumin, and globulin levels were all higher with plasma compared to serum. In addition, AST was also higher with plasma, but ALT was lower. The paper concluded that these differences were unlikely to be clinically important and that plasma provided reasonable approximations for serum biochemical values. Small blood tubes containing lithium heparin with plasma separators are especially useful in reptile practice as they come in a variety of sizes, as small as 400 μl.

Hemolysis is known to cause changes in biochemical results and clinicians should be vigilant to evaluate for any hemolysis before laboratory submission. In the green iguana, moderate hemolysis increases phosphorus levels, while marked hemolysis significantly increases phosphorus, potassium, total protein and AST.[13]

Samples should ideally be processed immediately; however, when delays of 1-48 hours are anticipated, plasma should be separated immediately from the cellular fraction, and kept refrigerated or, in field conditions, kept on ice. If delays exceed 2 days, then freezing at -70°C will preserve most chemical constituents.[14] Frozen-thawed plasma should be thoroughly mixed prior to processing.

References

1. Friedrichs KR, Harr KE, Freeman KP, et al. ASVCP reference interval guidelines: determination of de novo reference intervals in veterinary species and other related topics. *Vet Clin Pathol.* 2012;41(4):441-453.

2. Campbell TW. Clinical pathology. In: Mader DR, Divers SJ, eds. *Current Therapy in Reptile Medicine and Surgery.* 2nd ed. St Louis, MO: Elsevier;2014:70-92.

3. Campbell TW. Hematology of reptiles. In: Thrall MA, ed. *Veterinary Hematology and Clinical Chemistry.* London: Lippincott Williams and Wilkins;2004:259-276.

4. Dessauer HC. Blood chemistry of reptiles: physiological and evolutionary aspects. In: Gans C, Parsons TS, eds. *Biology of the Reptilia.* Vol 3. *Morphology C.* London: Academic Press;1970:1-72.

5. Girons MCS. Morphology of the circulating blood cells. In: Gans C, Parsons TS, eds. *Biology of the Reptilia.* Vol 3. *Morphology C.* London: Academic Press;1970:73-92.

6. Duguy R. Numbers of blood cells and their variation. In: Gans C, Parsons TS, eds. *Biology of the Reptilia.* Vol 3. *Morphology C.* London: Academic Press;1970:93-110.

7. Strik NI, Alleman AR, Harr KE. Circulating inflammatory cells. In: Jacobson ER, ed. *Infectious Diseases and Pathology of Reptiles - Color Atlas and Text.* London: Taylor and Francis;2007:167-218.

8. Campbell TW. Clinical chemistry of reptiles. In: Thrall MA, ed. *Veterinary Hematology and Clinical Chemistry.* London: Lippincott Williams and Wilkins;2004:493-498.

9. White FN. Circulation. In: Gans C, Dawson WR, eds. *Biology of the Reptilia.* Vol 5. *Physiology A.* London: Academic Press;1976:275-334.

10. Hanley CS, Hernandez-Divers SJ, Bush S, Latimer KS. Comparison of dipotassium EDTA and lithium heparin on hematologic values in the green iguana (*Iguana iguana*). *J Zoo Wildl Med.* 2004;35(3):328-332.

11. Mayer J, Knoll J, Innis C, Mitchell MA. Characterizing the hematologic and plasma chemistry profiles of captive Chinese water dragons, *Physignathus cocincinus. J Herpetol Med Surg.* 2005;15(3):45-52.

12. Perpinan D, Hernandez-Divers SM, Hernandez-Divers SJ. Comparison of three different techniques to produce blood smears from green iguanas, *Iguana iguana. J Herpetol Med Surg.* 2006;16(3):99-101.

13. Finkelstein AD, DiGesualdo CL, Hoover JP. Comparison of biochemical profile values from paired plasma and serum samples of green iguanas, *Iguana iguana*, and red-tailed boa constrictors, *Boa constrictor. J Herpetol Med Surg.* 2003;13(3):5-9.

14. Weiser G. Sample collection, processing, and analysis of laboratory service options. In: Thrall MA, ed. *Veterinary Hematology and Clinical Chemistry.* London: Lippincott Williams and Wilkins;2004:39-44.

Preventive Health Habits of Effective Herpetoculturists

Christina Miller RVT, BSc

Session #232

Affiliation: From Companion Animal Hospital, Dartmouth, Canada.

Roundtable Discussion Points: ExoticsCon 2016

1. Redefining success in herpetoculture: Cultural value of reproductive success over longevity and low incidence of preventable diseases. "Big box breeding" and concerns for individual health.

2. Teaching your clients to manage their collection like a zoo:

- Record-keeping for trend recognition
- Establishing a quarantine protocol
- Taking parasites seriously
- Building a good relationship with your vet
- Dealing with death in the collection; the value of necropsies
- Risk management in husbandry techniques

Section 28

ARAV Clinical Reviews and Care in Captivity

Dan Loper, DVM;
Kim Wojick, DVM;
Karina Mathes, DVM, DECZM (Herpetology);
Shane Simpson, BVSc (hons), GCM (VP), CMAVA;
Michael Miller, DVM
Moderators

A Non-Endoscopic Technique for Prefemoral Castration of Hermann's Tortoises

Nicola Di Girolamo, DMV, GPCert(ExAP), MSc(EBHC), PhD, Dipl ECZM (Herpetology)

Session #212

Affiliation: From the Clinica per Animali Esotici, Centro Veterinario Specialistico, Via S. Giovannini 53, 00137, Roma, Italy.

Castration may be indicated for ecological or behavioral reasons. Recently, an endoscopic technique has been described for castration of several chelonian species,[1] including Hermann's tortoises.[2] The endoscopic orchiectomy requires specific endoscopic skills and instrumentation. The purpose of this work is to describe a non-endoscopic technique for castration of Hermann's tortoises (*Testudo hermanni*). The technique was performed in six clinically healthy privately owned sub-adult Hermann's tortoises (body weight 240-410 grams) presented for population control. All the surgeries were performed at the beginning of the summer, i.e., sexual active season. A standard left prefemoral coeliotomy with the tortoise in right lateral recumbency was performed. Stay sutures were placed on 4 orthogonal points of the coelomic membrane to allow maximal retraction. The left testicle and epididymis were easily visualized. A hemostat was placed on the cranial side of the mesorchium. The testicle was gently moved cranially while dissecting the mesorchium from the epididymis and testis with a bipolar radiosurgical scalpel. After excision of the left testis, the prefemoral incision was closed in two layers, and the same procedure was performed for the left testis with the tortoise placed in left lateral recumbency. The surgeries lasted approximately 1 hour. Of the 12 testes removed, 1 ruptured from excessive pressure while being grasped and 1 experienced minimal bleeding that resolved with placement of compressive gauze. All animals recovered uneventfully and were alive 4 months postoperatively. The main limitation of this technique is the need to perform a bilateral celiotomy.

References

1. Innis CJ, Feinsod R, Hanlon J, Stahl S, Oguni J, Boone S, Schnellbacher R, Cavin J, Divers SJ. Coelioscopic orchiectomy can be effectively and safely accomplished in chelonians. *Vet Rec*. 2013;172(20):526.

2. Paries S, Funcke S, Ziegler L, Lierz M, Wüst E. Endoscopic assisted orchiectomy in Herman's tortoises (*Testudo hermanni* sp.). *Tierarztl Prax Ausg K Kleintiere Heimtiere*. 2014;42(6):383-389.

Effects of Plant Protein Diets on the Health of Farmed American Alligators *(Alligator mississippiensis)*

Peter M. DiGeronimo, VMD,
Nicola Di Girolamo, DMV, MSc,
Nicholas A. Crossland, DVM,
Fabio Del Piero, DVM, PhD, Dipl ACVP,
Robert C. Reigh, PhD, and
Javier G. Nevarez, DVM, PhD, Dipl ACZM, Dipl ECZM

Session #126

Affilitation: From Departments of Veterinary Clinical Sciences (DiGeronimo, Nevarez) and the Department of Pathobiological Sciences (Crossland, Del Piero), Louisiana State University School of Veterinary Medicine, Skip Bertman Dr, Baton Rouge, LA 70803, USA, la Clinica per Animali Esotici, Centro Veterinario Specialistico (CVS), Via Sandro Giovannini 53, 00137 Rome, Italy (Di Girolamo) and the Aquaculture Research Station, Louisiana State University Agricultural Center, 2410 Ben Hur Rd, Baton Rouge, LA 70820, USA (Reigh).

The objective of this prospective, blinded study was to compare plasma biochemical values and gross and histologic evaluation of kidney and liver from American alligators (*Alligator mississippiensis*) fed extruded diets with protein derived from animal or plant sources. Alligators in 2 treatment groups were fed an extruded diet with protein derived primarily from plant products for 7 (n = 20) or 10 (n = 20) months prior to harvest. A control group (n = 20) was fed a commercial diet with protein derived from animal products for the duration of the study. Plasma biochemistry panels, gross and histologic examination of kidney and liver tissues were obtained for each animal. No differences were found between alligators fed diets with animal or plant protein in either biochemistry profiles, or gross or histologic examination of kidney and liver. Plant-based diets, fed for up to 10 months, do not appear to have any ill effects on the kidney or liver of American alligators.

Carapacial Scute Pyramiding Increased by Heat in African Leopard and Spurred Tortoises

Mark L. Heinrich MS, DVM, and
Kaleb K. Heinrich, MS, DA

Session #100

Affiliation: From Carlsbad Animal Clinic, Living Desert Zoo and Gardens State park, 103 E Blodgett, Carlsbad, NM 88220, USA (M Heinrich) and Department of Biology, University of Mary Hardin Baylor, 900 College St, Belton, TX 76513, USA (K Heinrich).

Abstract: Carapacial scute pyramiding (CSP) is a common phenomenon in captive raised chelonians. Several hypothetical causes have been proposed but few have been scientifically investigated. We examined the effect of nocturnally applied heat in captive-hatched African leopard (*Stigmochelyes pardalis*) and spurred tortoises (*Centrochelyes sulcata*) over a 2-year period.

Monthly measurements of individual African leopard (*Stigmochelyes pardalis*) and spurred tortoises' (*Centrochelyes sulcata*) surface temperature, length, height, width, weight, and pyramid height of treatment (heat) and control (no heat) groups were taken over a 2-year period. These measurements were analyzed, as were air temperature and relative humidity data.

Daily environmental conditions and offered diet did not differ between treatment and control groups. Treatment and control groups were placed outside with their cohorts during daylight hours. If weather did not permit, they were exposed to the same thermostatically controlled air temperature and photoperiod inside. Therefore the treatment differed from control via the application of nighttime heat.

During the study all the tortoises' morphometric measurements increased but values were higher in treatment groups of both species at the end of the study. Furthermore, allometric comparison between control and treatment animals when they were of similar size revealed increased pyramiding in the treatment group.

Our results indicate that growth rate and CSP appear to be directly related and both increase with excess nocturnal heat. Normal growth in turtles and tortoises is ultimately dependent on an individual species' evolutionary history. As we learn more about a given species' natural history we will grow closer to providing environmental factors that will be favorable for normal growth in captivity.[1]

References

1. Heinrich ML, Heinrich KK. Effect of supplemental heat in captive African leopard tortoises (*Sigmocheyles pardalis*) and spurred tortoises (*Centrocheyles sulcata*) on growth rate and carapacial scute pyramiding. *J Exotic Pet Med.* 2016;25:18-25.

Squamous and Basal Cell Carcinoma in Reptiles

Tom Hellebuyck, DVM, PhD, Dipl ECZM (Herpetology),
Richard Ducatelle, DVM, PhD, Dipl ECVP,
Annemie Van Caelenberg, DVM, PhD,
An Martel, DVM, PhD, Dipl ECZM (Wildlife Population Health)

Session #177

Affiliation: From the Department of Pathology, Bacteriology and Avian Diseases (Hellebuyck, Ducatelle, Martel) and Department of Medical Imaging, Faculty of Veterinary Medicine (Van Caeleberg), Ghent University. Salisburylaan 133, B-9820 Merelbeke, Belgium.

The clinical behavior and histopathologic characteristics of an oral keratinizing basal cell carcinoma (BCC) and a metastatic carapacial BCC in two Hermann's tortoises (*Testudo hermanni*) were compared to various cases of squamous cell carcinoma (SCC) diagnosed in squamates and chelonians. To the best of our knowledge, the occurrence of BCC has not previously been documented in a reptilian species. While human cutaneous BCC is a highly common malignant neoplastic disorder, intraoral BCC in humans and BCC in domestic animals and birds are rare.[1-3] Exposure to ultraviolet (UV) radiation is the major cause of cutaneous BCC in humans.[1] UV radiation may also be a predisposing factor in the development of SCC and BCC in reptiles.[4] Every presumed case of SCC in reptiles should be differentiated from BCC as the predilection sites and gross pathologic features for both neoplasms are similar. In comparison to the observed SCCs, the BCCs showed a remarkably fast and highly infiltrative growth. Accordingly, an early diagnosis including reliable discrimination from SCC is essential towards the management of this neoplastic entity. SCC and BCC are curable conditions if a prompt diagnosis is accomplished and treatment is initiated in the early phase. When surgical excision is impossible, intralesional treatment may be considered for the treatment of dermal and intraoral SCC and BCC. The effect of intralesional treatment with 5-flourouracil was evaluated in the Hermann's tortoise with an intraoral BCC but did not prove to be effective.

References

1. Crowson AN. Basal cell carcinoma: biology, morphology and clinical implications. *Modern Pathol*. 2006;19:S127-147.

2. Campos AG, Cogliati B, Guerra JM, Matera JM. Multiple trichoblastomas in a dog. *Vet Dermatol*. 2014;25(1):48-e19.

3. Tell LA, Woods L, Mathews KG. Basal cell carcinoma in a blue-fronted Amazon parrot (*Amazona aestiva*). *Avian Dis*. 1997;41:755-759.

4. Hannon DE, Garner MM, Reavill DR. Squamous cell carcinomas in inland bearded dragons (*Pogona vitticeps*). *J Herpetol Med Surg*. 2011;21(4):101-106.

Shell Disease in Western Pond Turtles *(Actinemys marmorata)* in the State of Washington

Bethany Groves, DVM,
Katherine Haman, DVM, MSc,
Drury R. Reavill, DVM, Dipl ABVP (Avian), Dipl ABVP (Reptile and Amphibian), Dipl ACVP,
and Western Pond Turtle Health Team (PAWS, Woodland Park Zoo, Oregon Zoo, University of Illinois, Washington State University, WDFW staff)

Session #249

Affiliation: PAWS Wildlife Center, 15305 44th Ave W, Lynnwood, WA 98087, USA (Groves), Department of Fish and Wildlife, 1111 Washington Street SE, Olympia, WA 98501, USA (Haman), Zoo/Exotic Pathology Service, 6020 Rutland Drive #14, Carmichael, CA 95608, USA (Reavill).

The western pond turtle was listed as endangered in Washington in 1993, at which time recovery efforts were already underway. The recovery program has largely been a success and resulted in an overall population increase from 150 animals to more than 1000 individuals at six locations in Washington. However, recent observations identified shell disease of unknown etiology affecting turtles in all six populations in Washington. Retrospective review of photographs suggests shell disease was present in western pond turtle populations as early as 2003, when photo documentation was initiated. In some individuals, the disease is severe and these turtles require veterinary care, including surgical debridement and long-term antimicrobial therapy. The use of diagnostic imaging (computed tomography) has recently enhanced our ability to document the extent and severity of shell disease and focus treatment efforts. Further, diagnostic imaging has been used to document subclinical cases; to date, 3 out of 5 apparently healthy turtles were shown to have moderate to severe shell disease. Though the sample size is small, that is a potentially staggering subclinical prevalence. Molecular diagnostics are currently underway to characterize possible primary pathogens and identify shifts in the shell microbiomes associated with shell disease status.

Composition of Cloacal Bacterial Flora in Healthy *Testudo hermanni* and During Cloacitis

Emanuele Lubian, DVM, GPCert(ExAP),
Piera Anna Martino,
Massimo Millefanti, DVM

Session #168

Affiliation: From Ambulatorio veterinario, Via Galvani 42, Gaggiano, 20083, Italy (Lubian, Millefanti), and Università degli Studi di Milano, Dipartimento di Medicina Veterinaria (DIMEVET), microbiologia e immunologia, Via Celoria 10, Milano 20133, Italy (Marino).

Abstract: This study aims to standardize the bacterial cloacal composition in healthy Hermann's tortoise and to see how it changes during different seasons, in different places, with different sampling methods and during cloacitis. Samples were taken from selected tortoises four times during 2009. In the same year, cloacitis samples were taken from sick tortoises examined during clinical practice.

This study aims to standardize the bacterial cloacal composition in healthy Hermann's tortoise and to see how it changes during different seasons, in different places, with different sampling methods and during cloacitis. Samples were taken from 10 selected tortoises four times (April, July, September, October) during 2009 in 2 captive-bred groups. In the same year, cloacitis samples were taken from sick tortoises examined during clinical practice. Samples were taken with cloacal swab and cloacal lavage and analyzed with bacterial culture.

The conclusions of this study are that some bacteria (i.e., *Bacillus* sp., *Enterococcus* sp., *E. coli*, *Klebsiella* sp.) could be found in healthy tortoises throughout the year. Other bacteria (i.e., *Pseudomonas* spp., *Pantoea* spp., *Salmonella* spp., *Proteus mirabilis*, *Acinetobacter* spp.) show higher prevalence during some seasons and show lower prevalence or are completely absent in other seasons. Cloacal bacteria concentration seems to be associated with animal metabolism, so it is higher during the summer season and lower pre- and post-hibernation.

Studying the differences between cloacal swab and lavage reveal that the sensibility of the two methods is similar but we found that flushing is a better method because it allowed us to have quantitative, in addition to qualitative, results.

During the year we examined 3 tortoises with cloacitis. The bacterium isolated was always *Pseudomonas aeruginosa*. In one case it was associated with *Proteus mirabilis*. As a result, we can say that cloacitis is characterized by overgrowth of normal cloacal bacteria, which are opportunistic agents.

References

1. Caldora C, Cocilovo A, Dall'Ara P, Martino PA. Classificazione dei batteri e batteriologia speciale. In: Poli G, Cocilovo A. *Microbiologia e Immunologia Veterinaria*. UTET, Torino, 2005;207.

2. Candela MG, Atance PM, Seva J, Pallarés FJ, León Vizcaíno L. Granulomatous hepatitis caused by *Salmonella typhimurium* in a spur-thighed tortoise (*Testudo graeca*). Vet Rec. 2005;157:236-237.

3. Dickinson VM, Duck T, Schwalbe CR, Jarchow JL, Trueblood MH. Nasal and cloacal bacteria in free-ranging desert tortoises from the western United States. *J Wildl Dis*. 2001;37(2):252-257.

4. Espinosa-Aviles D, Salomon-Soto VM, Morales-Martinez S. Hematology, blood chemistry, and bacteriology of free ranging Mexican beaded lizard (*Heloderma horridum*). *J Zoo Wildl Med*. 2008;39(1):21-27.

5. Hidalgo-Vila J, Dıaz-Paniagua C, De Frutos-Escobar C, Jimenez-Martınez C, Perez-Santigosa N. *Salmonella* in free living terrestrial and aquatic turtles. *Vet Microbiol*. 2007;119:311-315.

6. Sunderland R, Veal J. An investigation into the faecal flora of clinically healthy tortoises, *Testudo*. *J Br Chelonia Group*. 2001;5:23-34.

Food Intake in Leopard Geckoes *(Eublepharis macularius)* During the Reproductive Period

P. Koelle, Dr. med. vet., Priv.Doz., Cert. Spec. for Reptiles, Cert. Spec. for Fish (including Ornamental Fish), Cert. Spec. for Nutrition Consultation (Pets)

Session #142

Affiliation: From the Clinic for Small Animal Internal Medicine of Ludwig-Maximilians-Unversity Munich, Veterinaerstrasse 13, D-80539 Muenchen, Germany.

Food intake of 8 adult leopard geckoes (3 male and 5 female specimens, 4-23 years old, average body weight: 45.2 g) was monitored during breeding season for a period of 30 days to investigate the maintenance feeding requirements of adult breeding leopard geckoes (*Eublepharis macularius*). The leopard geckoes were owned by a private keeper and kept in terraria with different temperature zones.

The animals were fed individually with forceps for the cockroaches (*Phoetalia pallida*) and kingworms (*Zophobas morio*) were fed ad libitum on average every 4-5 days. Weender analysis procedure of the feeder insects was performed. Every insect was weighed before feeding it to the leopard geckoes before and after dusting with vitamin and mineral supplement (Korvimin ZVT®). Intake of wet weight was recorded. From these results intake of wet weight, dry matter, and caloric intake was calculated for each leopard gecko.

On average each leopard gecko ate 0.53 g insects/day (wet weight) and 0.22 g dry matter/day. Caloric intake was calculated at 36.0 kcal gross energy per leopard gecko per month. In both sexes food intake was significantly higher during the breeding season. It was higher in male specimens in comparison to female ones due to feeding gaps of several days before laying eggs in female leopard geckoes.

References

1. Koelle P. Food intake in leopard geckoes (*Eublepharis macularius*). *Proc Annu Conf Assoc Rept Amph Vet.* 2015;29.

Water Quality for Herptiles: Not Just a Fishy Business

Ryan O'Shea, CVT, and Ray Ball, DVM

Session #045

Affiliation: From Tampa's Lowry Park Zoo Veterinary Department, 1101 W Sligh Ave, Tampa, FL 33604, USA.

Water quality analysis is the single most important aspect of collecting a thorough history of ornamental fish and a similar emphasis should be placed on water quality of aquatic herptiles. Parameters commonly assessed in aquaculture include, but are not limited to: pH, temperature, total ammonia nitrogen, dissolved oxygen, and chlorine; each of which may negatively affect the health of inhabitants when not properly maintained. Monitoring the water quality of aquatic herptiles is a large portion of the patient history and is needed to form a complete assessment of the patient.

Axolotls for Amateurs: A Case Study of Surgical Gastric Foreign Body Removal in a Juvenile Mexican Axolotl

Marcie Logsdon, DVM

Session #172

Affiliation: From the Washington State University College of Veterinary Medicine, 205 Ott Rd., Pullman WA 99164, USA.

Abstract: A juvenile Mexican axolotl (*Ambystoma mexicanum*) presented for ingestion of multiple pieces of aquarium gravel. The presence of gastric foreign bodies was confirmed with transillumination of the coelomic cavity and radiographs taken with a dental radiography unit. Surgical removal of the gastric foreign bodies was successful. Immersion induction and branchial anesthesia was provided with alfaxalone. Recovery was uneventful. Despite its small size and aquatic nature, surgeries under general anesthesia can be safely and effectively performed on this species with equipment available in most veterinary hospitals.

Introduction

The Mexican axolotl (*Ambystoma mexicanum*) is an aquatic salamander that was originally introduced to the United States of America for use in research settings. Its neotenic and exclusively aquatic life cycle and its ability to regrow severed limbs make it a useful laboratory animal for many studies. It is also gaining popularity as a pet and many color morphs, including a genetically modified strain that contains the green fluorescent protein (GFP) are sold as pets by hobby breeders. These fascinating animals do occasionally present to veterinary clinics for treatment. When they do, special considerations for housing, anesthesia and surgery must be taken into account. However, most veterinary clinics already have the necessary equipment, or the means to improvise it, on hand. This clinical report outlines the diagnosis and treatment of a common presenting complaint in axolotls: multiple foreign body ingestion. It also discusses realistic tips and techniques for managing axolotls in a hospital setting.

Clinical Report

In June 2015 an immature Mexican axolotl (*Ambystoma mexicanum*) presented to the Washington State University Veterinary Teaching Hospital Exotics Department for probable foreign body ingestion. The owner had observed the axolotl swallowing a piece of aquarium gravel during feeding time. On physical exam the axolotl weighed 24 g and was missing all but the approximately 0.5 cm of the gills on its right side. This was an old injury likely sustained from other hatchlings. Transillumination of the coelomic cavity with a Finoff transilluminator yielded a large opacity in the vicinity of the stomach. Dental radiographs revealed multiple gastric foreign bodies (Fig 1). The patient was housed in dechlorinated water in a 1 L Tupperware® container in the refrigerator overnight at approximately 40°F[1] in an attempt to decrease stress and possibly incite regurgitation of the foreign bodies.

The next morning a scant amount of feces was present, however, the axolotl had not regurgitated any of the gravel. Due to the small patient size, large number of foreign bodies, and high risk of esophageal or gastric perforation, endoscopic retrieval of the foreign bodies was not attempted and the decision was made to perform a gastrotomy.

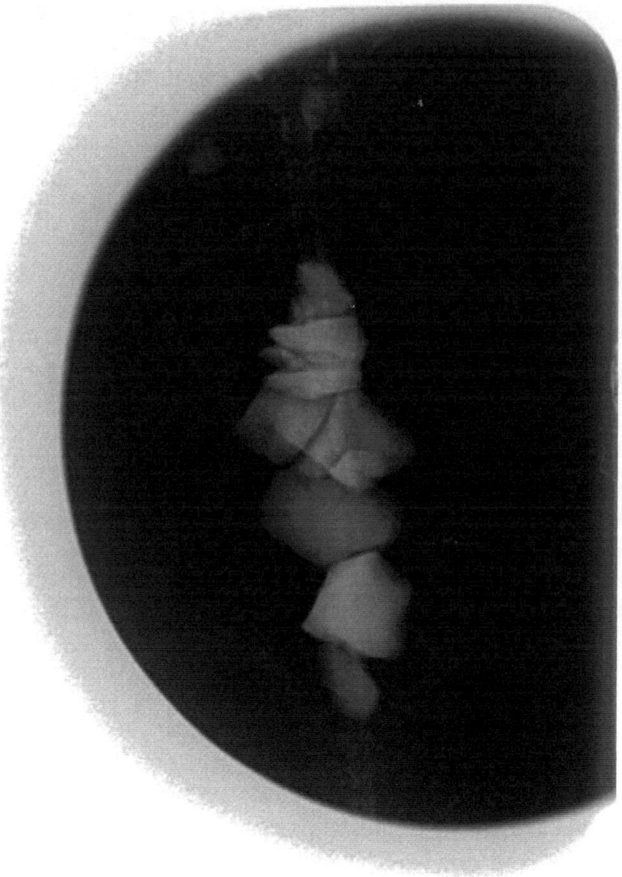

Figure 1. Axolotl radiographs taken with a dental unit showing multiple gastric foreign bodies.

Induction was performed by placing the axolotl into a small, 0.5 L container of his tank water containing alfaxalone (Alfaxan, Jurox Inc., Kansas City, MO, USA) at a concentration of 5 mg/L.[2] Additional 10 mg/ml alfaxalone (Jurox) was titrated to the water directly over his gills to effect. Once the axolotl exhibited loss of the righting reflex it was moved to an improvised surgical table. The surgical table was created by inverting the lid of a small Kritter Keeper® over its base and lining it with saline-soaked gauze. This provided a moist surgery surface that would allow drainage and collect used anesthetic solution throughout the procedure. The patient was placed in dorsal recumbency with two pieces of rolled, saline-soaked gauze to help hold him in position. During the procedure branchial anesthesia was performed by manually irrigating the gills with alternating 5 mg/L alfaxalone (Jurox) solution and sterile saline to effect. Sterile saline was also applied continuously over the body to prevent desiccation during the procedure. A 30-U insulin syringe of full-strength 10 mg/ml alfaxalone (Jurox) was used as needed to apply direct brachial anesthesia. Depth of anesthesia was monitored by visual assessment of respiratory rate (gill flutters), heart rate (readily visible through the body wall), and responsiveness. The surgical site was prepped with a single pass of dilute chlorhexidine solution. A modified plastic sterile sleeve was placed over the patient and attached to the gauze using towel clamps.

An approximately 1.5 cm skin incision was made over ventral midline. The coelomic musculature and lining were elevated using thumb forceps and a small stab incision was made into the coelomic cavity. The incision was extended cranially and caudally to match the skin incision. Stay sutures of 6-0 Polyglactin 910 (Vicryl,

Ethicon, Inc., Summersville, NJ, USA) were placed full-thickness to aid in retraction. A sterile microswab was moistened with sterile saline and bent into a hook to exteriorize the stomach. The stomach was packed off and isolated from the coelomic cavity using moistened, sterile swabs. Additional stay sutures of 6-0 Polyglactin 910 (Ethicon) were placed in the stomach as well. An approximately 1cm incision was made into the stomach and the ingested rocks were removed using DeBakey forceps. Two intraoperative dental radiographs (Figs 2 and 3) were taken to ensure that all rocks were removed. The entire surgical platform (Kritter Keeper® lid) was easily covered with a sterile plastic sheet and transported en bloc to the dental radiography unit while maintaining sterility of the surgical field. A total of 15 rocks varying in size from 3 mm to 1 cm were removed. The stomach was closed using 6-0 Polyglactin 910 (Ethicon) in a simple continuous pattern with a Lehmbert over sew. A single layer closure of the coelomic membrane, musculature, and skin was performed in a simple continuous pattern using 6-0 Polyglactin 910 (Ethicon).

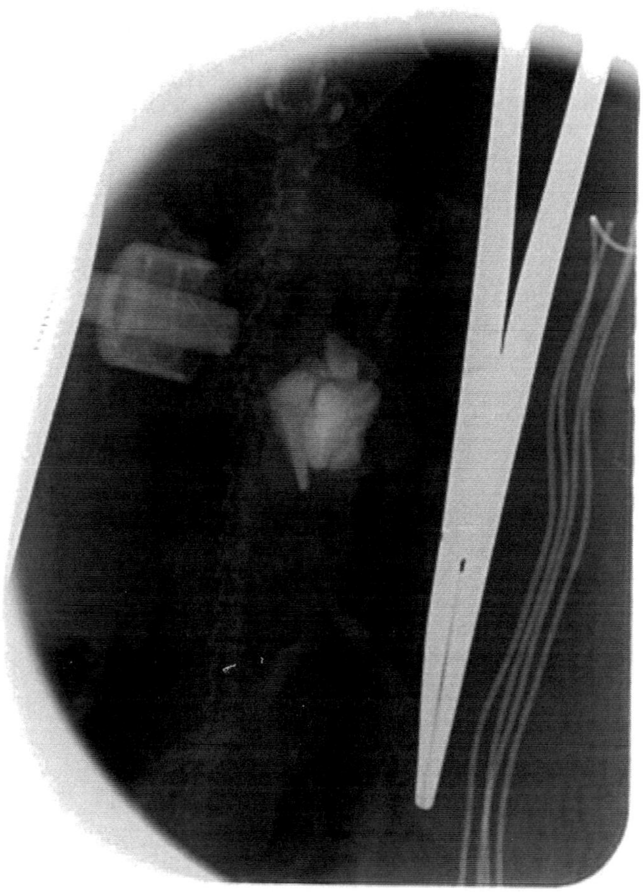

Figure 2. Intraoperative radiographs showing multiple foreign bodies remaining. The radiopaque strip in a piece of detectable gauze is visible as well as the line providing branchial anesthesia and a pair of hemostats that are in place holding stay sutures.

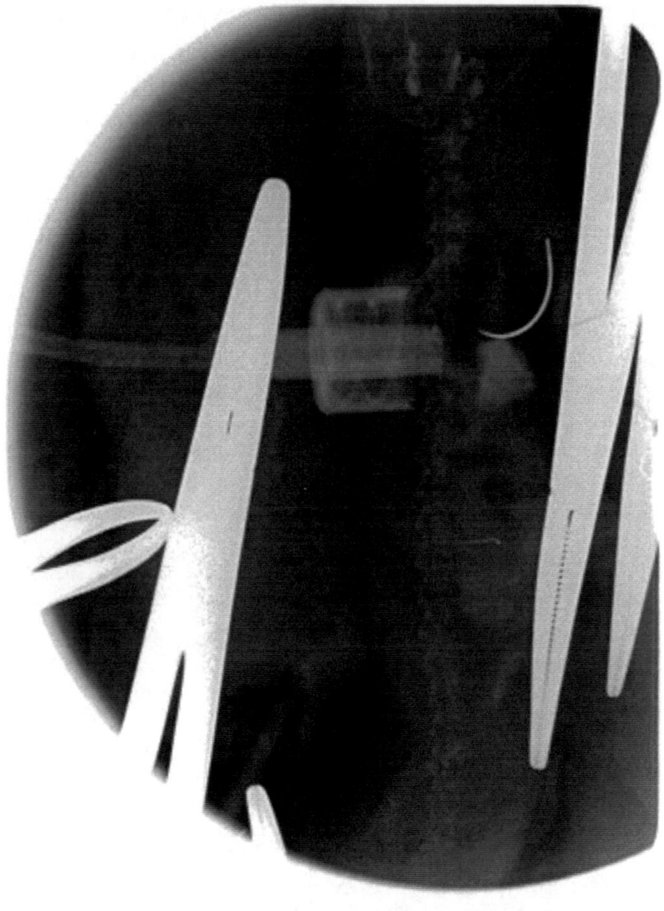

Figure 3. Final intraoperative radiograph showing single remaining gravel piece. Anesthesia line, multiple hemostats holding stay sutures, towel clamp, and swadged on suture needle overlying the radiographic field as well.

Post-operatively the patient was administered 0.23 mg enrofloxicin (Baytril, Bayer Healthcare LLC, Animal Health Division, Shawnee Mission, Kansas, USA) and 1.5 µg buprenorphine (buprenorphine hydrochloride, Par Pharmaceuticals, Cos. Inc., Spring Valley, NY) intramuscularly in the tail. The patient was recovered in dechlorinated water over a period of approximately 30 minutes.

After recovery the patient was maintained in a 1 L Tupperware® container housed in a mini-fridge with temperatures ranging from 40-50°F. Excluding the first day post-surgery, enrofloxicin (Baytril 100, Bayer Healthcare LLC, Animal Health Division, Shawnee Mission, Kansas, USA) baths[3] of 0.3 mg/L were performed once daily for approximately three hours for four days. On Day five the patient was moved to a 10-gallon aquarium containing dechlorinated water and was allowed to warm to its preferred optimum temperature range (POTR) of approximately 64-69°F.[4,5] While in hospital this temperature was maintained by twice daily rotation of cold packs that kept the temperature within the necessary parameters and also provided a gradient effect. The cold packs were made of 1 L sterile saline pour bottles filled with dechlorinated water. The patient's appetite returned shortly after cessation of fridging. The patient's incision healed without incident and the sutures were removed approximately 8 weeks post-surgery using manual restraint. This clinical report demonstrates the ability to provide an excellent standard of care for aquatic salamanders with equipment and supplies that are readily accessible to most practicing veterinarians.

References

1. Axolotl Sanctuary. Fridging. Available at: http://www.caudata.ort/axolotl-sanctuary/Fridging.shtml. Accessed May 8, 2016.

2. McMillan MW, Leece EA, Immersion and branchial/transcutaneous irrigation anaesthesia with alfaxalone in a Mexican axolotl. *Vet. Anaesth. Analg.* 2011;38:619-623.

3. Menger B, Vogt PM, Jacobsen ID, Allmeling C, et al. Resection of a large intra-abdominal tumor in the Mexican axolotl: a case report. *Vet. Surg.* 2010;39:232-233.

4. Reptiles Magazine. Axolotl Care Sheet. Available at http://www.reptilesmagazine.com/Care-Sheets/Frogs-Amphibians/Axolotl/index.php. Accessed June 4, 2015.

5. Axolotl.Org. Axolotls. Available at http://www.axolotl.org/. Accessed May 2, 2016.

Evaluation of Photobiomodulation in the Treatment of Epithelial Trauma in the Green Iguana *(Iguana iguana)*

Lara Cusack, DVM,
Joerg Mayer, DVM, MS,
Daniel Cutler, DVM,
Daniel R. Rissi, DVM, MS, PhD,
Stephen Divers, DZooMed

Session #194

Affiliation: From the University of Georgia, College of Veterinary Medicine, Department of Small Animal Medicine and Surgery, 2200 College Station Rd, Athens, GA 30602, USA.

The efficacy of photobiomodulation and topical therapy in the treatment of epithelial trauma in the green iguana (*Iguana iguana*) was evaluated. Sixteen healthy, sub adult green iguanas were used for the study. Under anesthesia, 3 5-mm epithelial biopsies were taken from each animal. Animals were divided into 2 treatment groups. Topical therapy wounds received either silver sulfadiazine (SSD), Tricide™, or no treatment. Laser therapy wounds received treatment with a class 4 laser at either 5 J/cm^2, 10 J/cm^2, or no treatment. Wound measurements were obtained daily for 14 days. Animals were euthanatized and the treatment sites were evaluated microscopically for ulceration, inflammation, fibrosis, presence of bacteria, and collagen maturity. At the time of submission, final statistical analysis was not completed. Final results to be discussed at the time of presentation.

Effects of a Novel Light Emitting Diode on the Production of 25-hydroxycholecalciferol in the Bearded Dragon *(Pogona vitticeps)*

Lara Cusack, DVM,
Sam Rivera, DVM, MS, Dipl ABVP (Avian), Dipl ACZM,
Brad Lock, DVM, Dipl ACZM,
Stephen Divers, BVetMed DZooMed, Dipl ECZM (Herpetology),
Dipl ECZM (Zoo Health Management), Dipl ACZM, FRCVS

Session #197

Affiliation: From the University of Georgia College of Veterinary Medicine, Department of Small Animal Medicine and Surgery, 2200 College Station Rd, Athens, GA, 30602, USA (Cusack, Divers), and the Department of Animal Health, Zoo Atlanta, 800 Cherokee Avenue SE, Atlanta, GA 30315-1440, USA (Rivera, Lock).

The requirement for vitamin D has been documented in multiple reptile species, with deficiencies resulting in alterations in calcium homeostasis, nutritional secondary hyperparathyroidism, and metabolic bone disease. Though vitamin D3 can be obtained directly from dietary sources or from photobiosynthetic production, there exists species variability in the absorption and utilization of oral vitamin D, making exposure to ultraviolet B radiation an essential requirement for some species. The effect of different lights to promote synthesis of 25-hydroxycholecalciferol (25(OH)D) and cholecalciferol in the bearded dragon (*Pogona vitticeps*) was evaluated. Groups containing 5 animals each were exposed to industry standard fluorescent UVB lights, no UVB lights, and a prototype LED UVB light for a period of 11 months. Weekly measurement of UV output of the lights was performed. Cholecalciferol, 25(OH)D, ionized calcium, total calcium, and phosphorus were measured at time zero, at four months, eight months, and 11 months. Radiographs were taken at each time point to evaluate bone density. Parameters were measured between groups and time points. At the time of submission, final statistical analysis was not completed. Final results will be discussed at the time of presentation.

Section 29

ARAV Therapeutics and Case Studies

Erica Giles, DVM;
Jeff Baier, DVM;
Thomas Torres, DVM;
Stacey Leonatti, DVM, DABVP (Reptile & Amphibian)
Moderators

Leukemic Multicentric Lymphoma in a Green Tree Python *(Morelia viridis)* Treated with Prednisolone and L-Asparaginase

Emanuele Lubian, DVM, GPCert(ExAP),
Diana Binanti, DVM, PhD, Dipl ECVP,
Vincenzo Montinaro, DVM,
Massimo Millefanti, DVM

Session #165

Affiliation: From Ambulatorio veterinario, Via Galvani 42, Gaggiano, 20083, Italy (Lubian, Millefanti), AbLab veterinary diagnostic laboratory, Via Privata Massa Neri 13, Sarzana 19038, Italy (Binanti), Clinica veterinaria nervianese SRL, Via Lampugnani 3, Nerviano, 20014, Italy (Montinaro), Clinica Veterinaria Città di Vigevano, Viale dei Mille 22, Vigevano, 27029 Italy (Lubian).

Abstract: This case report describes a clinical approach, therapy, macroscopic and microscopic examination in a case of leukemic lymphoma with a primary thymic mass. The snake showed a swelling in the cardiac area. A diagnosis of lymphoma was made using ultrasound and cytologic examination. A complete blood count revealed high WBC count suggestive of leukemia. Therapy was planned with prednisolone (2 mg/kg IM q24h) and L-asparaginase (400 U/kg IM in single administration). Before and during therapy the dimension of the primary mass was measured every 4-7 days. After 1 month the snake showed a cloacal prolapse and a stomatitis, so treatment with enrofloxacin (10 mg/kg IM q24h) was started. After 2 months of therapy the animal died and the postmortem showed the presence of neoplastic cells in the kidneys, spleen, trachea, lung, esophagus, stomach, intestine, bone marrow, skin and skeletal muscle.

Introduction

In reptile practice, oncology represents a new field yet to be explored, although neoplasia is commonly encountered in these animals, with the highest prevalence in snakes. The present report describes a clinical approach, therapy, macroscopic and microscopic examination in a case of leukemic lymphoma with primary thymic mass.

Case Report

A 4-year-old, 620-g, captive bred, female green tree python (*Morelia viridis*) was examined for a swelling in the cardiac area. A diagnosis of lymphoma was made using ultrasound and cytologic examination of the mass. Blood cell count revealed high WBC count suggestive of leukemia. Therapy was planned with prednisolone (2 mg/kg IM q24h) and L-asparaginase (400 U/kg IM in single administration). Before and during therapy the dimension of the primary mass was measured every 4-7 days and it decreased a bit. After 1 month the snake showed a cloacal prolapse and a stomatitis, so treatment with enrofloxacin (10 mg/kg IM q24h) was started. After 2 months of therapy the animal died and the postmortem showed the presence of neoplastic cells in the kidneys, spleen, trachea, lung, esophagus, stomach, intestine, bone marrow, skin and skeletal muscle.

Neoplastic lesions are uncommon in snakes. Among them, lymphoma is the most commonly reported; however, description of possible therapeutic approaches are extremely rare in the veterinary literature. This is the first report of lymphoma in a green tree python (*Morelia viridis*) and the use of the drugs described in this case of lymphoma in a snake.

References

1. Schilliger L, Rossfelder A, Bonwitt J, Di Girolamo N, et al. Antemortem diagnosis of multicentric lymphoblastic lymphoma, lymphoid leukemia, and inclusion body disease in a Boa constrictor (*Boa constrictor imperator*). *J Herpetol Med Surg*. 2014;24(1-2):1-9.

2. Folland DW, Johnston MS, Thamm DH, Reavill D. Diagnosis and management of lymphoma in a green iguana (*Iguana iguana*). *J Am Vet Med Assoc*. 2011;239(7):985-991.

3. Mauldin GN, Done LB. Oncology. In: Mader DR, ed. *Reptile Medicine and Surgery*. 2nd ed. St Louis, MO: Elsevier Saunders;2006:299-322.

4. Georoff TA, Stacy NI, Newton AN, et al. Diagnosis and treatment of chronic T-lymphocytic leukemia in a green tree monitor (*Varanus prasinus*). *J Herpetol Med Surg*. 2009;19:106–114.

5. Lawton MPC. 1999. Pain management after surgery. *Proc Annu Conf North Am Vet Comm*. 782.

6. Stein G. Reptile and amphibian formulary. In: Mader DR, ed. *Reptile Medicine and Surgery*. St. Louis, MO: Elsevier Saunders;2006:465-472.

7. Sykes JM, Trupkiewicz JG. 2006. Reptile neoplasia at the philadephia zoological garden, 1901-2002. *J Zoo Wildl Med*. 37(1):11-19.

Devriesea agamarum Affects Endangered Lesser Antillean Iguanas *(Iguana delicatissima)*

Tom Hellebuyck, DVM, PhD, Dipl ECZM (Herpetology),
Karl Questel,
Frank Pasmans, DVM, PhD, Dipl ECZM (Herpetology),
Leen Van Brantegem, DVM, Dipl ECVP,
Pascal Philip,
An Martel, DVM, PhD, Dipl ECZM (Wildlife Population Health)

Session #176

Affiliation: From Department of Pathology, Bacteriology and Avian Diseases, Faculty of Veterinary Medicine, Ghent University, Salisburylaan 133, B-9820 Merelbeke, Belgium (Hellebuyck, Pasmans, Van Brantegem, Martel) and Agence Territoriale de l'Environnement de Saint-Barthélemy, PO Box 683, Gustavia, Saint Barthélemy (Questel, Philip).

Since April 2011 dermatologic disease has been observed in free-ranging, endangered Lesser Antillean iguanas (*Iguana delicatissima*) on the French Caribbean island of Saint Barthélemy. Lesions mainly consist of multiple large cutaneous and subcutaneous abscesses. Cloacal and oral samples collected in 10 Anguilla bank ameivas (*Ameiva plei*), 5 Anguilla bank anoles (*Anolis gingivinus*), 16 *I. delicatissima* and one *I. delicatissima* × green iguana (*I. iguana*) hybrid were submitted for microbiologic culturing. In 8 out of the 16 sampled *I. delicatissima* dermal lesions were observed. From these iguanas swabs and full thickness skin biopsies were collected from the lesions for microbiologic and histopathologic examination. *Devriesea agamarum* was isolated from the skin lesions in all affected iguanas, the oral cavity of 1 affected iguana and the cloaca of 1 healthy iguana and 2 healthy anoles. The histologic features of the affected skin sites where identical to those previously described for *D. agamarum* associated dermatitis and abscesses in captive lizards.[1] The results of this study provide strong evidence for a primary etiologic role of *D. agamarum* in the development of skin disease in free-ranging Lesser Antillean iguanas. Moreover, healthy lizards were identified as asymptomatic reservoirs. The emergence of *D. agamarum* in this population of *I. delicatissima* constitutes a new threat for this endangered lizard species and is the first description of a bacterial disease that poses a conservation threat towards free-ranging squamates.

References

1. Hellebuyck T, Martel A, Chiers K, Haesebrouck F, et al. *Devriesea agamarum* causes dermatitis in bearded dragons (*Pogona vitticeps*). *Vet Microbiol*. 2009;134:267-71.

Use of Voriconzole in Treatment of *Chrysosporium* spp. in a Bearded Dragon *(Pogona vitticeps)*

Rachel Williams, DVM

Session #038

Affiliation: From Webster Groves Animal Hospital, 8028 Big Bend Blvd, St. Louis, MO 63119, USA.

Abstract: Two female bearded dragons (*Pogonas vitticeps*) were presented for dysecdysis. On examination, bearded dragon 1 had an epidermal lesion on the mandible and the other (bearded dragon 2) on the dorsum. The lesions comprised large areas of flaking scales with purulent debris underneath. Cytology revealed mixed inflammatory cells and possible fungal hyphae. When no improvement was seen with Enterderm topical ointment, swabs of the lesions were submitted for culture, and cloacal and lesional swabs were submitted for viral and *Chrysosporium* anamorph of *Nannizziopsis vriesii* (CANV) PCR testing (respectively). Bearded dragon 1 was positive for *Adenovirus* and negative for CANV. The culture showed mild growth of *Staphylococcus*, and potential *Trichophyton* spp. on the fungal culture. This dragon's condition deteriorated and it was euthanatized. Bearded dragon 2 was negative for both *Adenovirus* and CANV. Its culture showed moderate to heavy growth of *Pseudomonas aeruginosa* and *Citrobacter*. Skin biopsy revealed fungal hyphae and structures reminiscent of arthrospores, considered consistent with CANV, or yellow fungus disease. Treatment with voriconazole was commenced at 10 mg/kg q24h and the dragon was deemed to be clinically normal after approximately 130 days.

CASE 1

An adult female bearded dragon (*Pogona vitticeps*) of unknown age, which was purchased from a pet store, was presented for evaluation of epidermal lesions on the mandible. The owner had noticed small skin lesions on the face at the time of purchase that started spreading approximately 6 months after purchase. The dragon was housed in a 4 × 2 × 2 foot wooden enclosure with a plexiglass front and a screen top. The enclosure also had a 150 watt UVB bulb (Exo terra Solar Glo Mercury Vapor) and a 125 watt infrared heat lamp (Westinghouse R40) used only in the winter or if the room temperature was below 21.1°C. The enclosure was maintained between 24-43°C during the day and 18°C at night. The diet consisted of primarily collard greens and various lettuces, blueberries, Dubia roaches with supplemental calcium powder and reptile vitamins.

The initial physical examination revealed necrosis and loss of some scales on the right side of the mandible with purulent discharge underneath. The dragon was determined to be underweight at the time. Cytology of the lesions revealed red blood cells, inflammatory cells, and probable fungal organisms (Figs 1 and 2). Topical Enterderm (nystatin-neomycin sulfate-thiostrepton-triamcinolone acetonide ointment, MWI, Boise, ID) was instituted once daily. The lesions did not respond to this course of treatment.

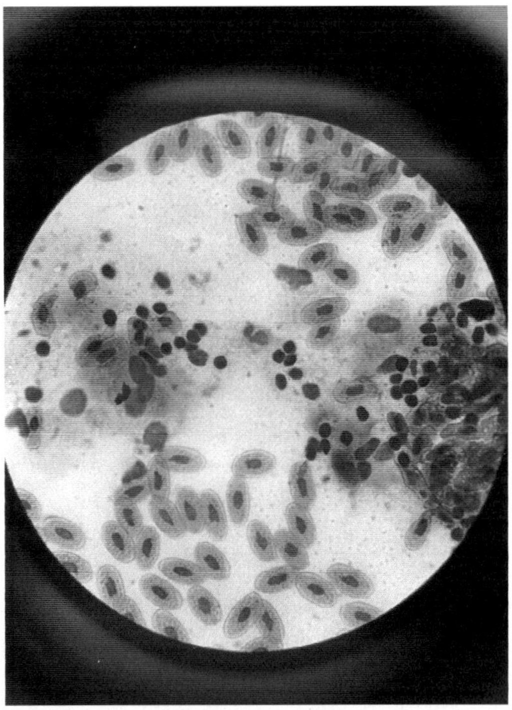

Figure 1. Cytology (image courtesy of Allen Weltig, DVM; 100x magnification).

The dragon, over approximately 5 weeks' time, became progressively anorexic and the lesions spread to the ventral inguinal area and dorsal aspect of the left hind foot. In light of the progressive course of the lesions, cloacal swab PCR testing was sent out to Avian and Exotics Clinical Pathology Lab for *Adenovirus*, swabs of the lesions were sent for PCR CANV testing, as well as bacterial and fungal cultures of the mandibular lesion. Results of the adenovirus PCR were positive, the CANV DNA PCR was negative, and culture of the mandibular lesion showed moderate fungal growth that was described as probable *Trichophyton* species. The Gram's stain of the lesion revealed 100% gram-positive cocci.

The dragon developed complete anorexia and diarrhea over the next week despite treatment with Entederm and syringe feeding with Carnivore Care (Oxbow Animal Health, Murdock, NE). The owner elected to humanely euthanatize her due to poor prognosis. Necropsy was not performed.

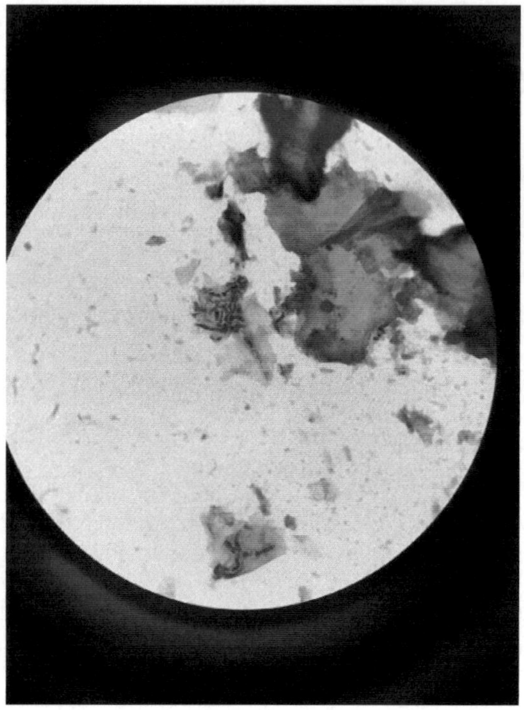

Figure 2. Cytology showing probable fungal hyphae (image courtesy of Allen Weltig, DVM; 100x magnification).

CASE 2

An adult female bearded dragon (*Pogona vitticeps*) of unknown age was purchased at the same time and from the same source as CASE 1. The dragon was housed in the same enclosure with bearded dragon 1, and fed the same diet. CASE 2 presented at the same time as CASE 1 for evaluation of a small dermal lesion affecting approximately 10% of the skin on the right side of her back. The dragon's attitude and appetite were normal at the time and she weighed 450 g. Over the next 4-5 months, scale loss, darkening and crusting of the skin, and spreading of the lesion to approximately 50% of the right side of her back occurred. Purulent discharge was noted underneath the scales that remained (Fig 3). Cytology of the lesion revealed the same as Case 1, and topical Entederm once daily was instituted.

Figure 3. Case 2 at time of biopsy (image courtesy of Alicia Yerkes).

The lesion continued to worsen over the next 5 weeks, although the patient's general demeanor and appetite remained the same. Cloacal swab PCR's were submitted for *Adenovirus* testing, and swabs of the lesions were submitted for CANV DNA to Veterinary Molecular Diagnostics (Milfod, OH), as well as bacterial and fungal cultures of the dorsal lesion to Avian and Exotics Clinical Pathology Laboratory. Both PCR tests were negative. The bacterial culture showed moderate-heavy growth of *Pseudomaonas aeruginosa*, and moderate growth of *Citrobacter*. The fungal culture was negative. A Gram's stain of the same lesion showed 90% gram-positive cocci, 1% gram-negative rods, and 9% gram-positive rods, and bacterial growth was listed as heavy by the laboratory.

Due to the progressive deteriorating nature of the lesion, lack of response to treatment, culture results and clinical suspicions, a skin biopsy under sedation was elected. (Fig 4) The dragon was sedated with a combination of intramuscular butorphanol, (Dolorex, Merck Animal Health, Summit, NJ) midazolam (Akorn, Inc., Lake Forest, IL), ketamine (Zetamine, MWI, Boise, ID) and dexmeditomidine (Zoetis, Kalamazoo, MI). A full thickness skin biopsy was taken and sent in formalin to Antech Clinical Pathology Laboratory (Irvine, CA). Results revealed fungal hyphae and structures reminiscent of arthrospores considered consistent with CANV, or yellow fungus disease.

Figure 4. Skin biopsy (image courtesy of Rachel Williams, DVM).

In light of this finding, a treatment trial with voriconazole was commenced. The selection of voriconazole was based on a previous study in bearded dragons by Waeyenberghe et. al.[1] In this study, voriconazole at 10 mg/kg had the most positive clinical outcome when compared with itraconazole. Two 200-milligram voriconazole tablets were obtained from online pharmacy Health Warehouse (healthwarehouse.com, Florence, KY) and compounded to 40 mg/ml in sugar syrup. Dosing regimen was instituted at 10 mg/kg orally q24h for approximately 50 days. The patient's skin was looking clinically much improved at the end of this treatment regimen. Length of treatment was based on clinical response of the patient to the treatment and previously published clinical trials (Fig 5).

Figure 5. Forty-nine days into the first round of treatment with voriconazole (image courtesy of Alicia Yerkes).

Approximately 14 days after the end of the course of treatment, a repeat full thickness skin biopsy was taken and sent in formalin to Antech laboratory. While initial histology results were negative for any presence of fungal organisms, the fungal staining showed a very low presence of fungal organisms consistent with CANV. Treatment with voriconazole was re-instituted for another 50 days at the same dose.

At approximately 15 days into the second course of treatment, it was noted that the lesion appeared to be darkening in color and it was feared that the disease was worsening (Fig 6). Eventually the entire lesion flaked off, revealing dry normal scales underneath. The treatment was discontinued at approximately 50 days, and the dragon was reexamined approximately 3 weeks after discontinuation of the treatment. At the time of examination, the skin was clinically normal (Fig 7), and further histopathology was declined by the client.

Figure 6. Fifteen days into the second round of treatment with voriconazole (image courtesy of Alicia Yerkes).

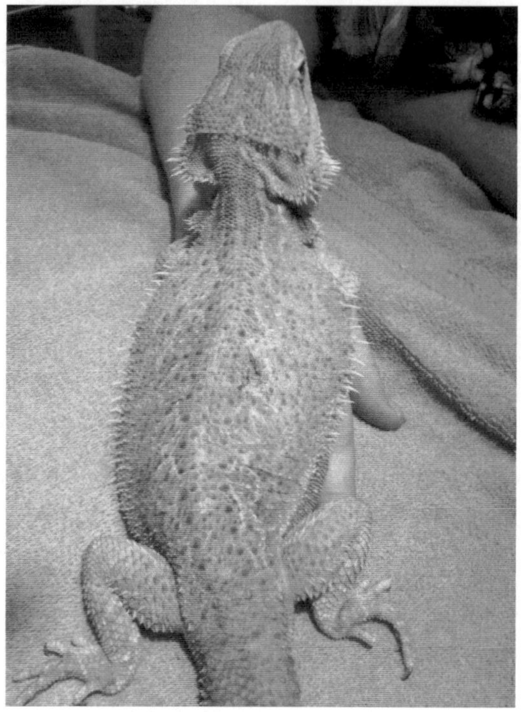

Figure 7. Approximately 3 weeks after cessation of the second round of treatment (image courtesy of Rachel Williams, DVM).

Discussion

Chrysosporium spp. are a large group of contagious fungal pathogens that are uncommonly implicated in cases of severe and life-threatening dermatitis in multiple orders of wild and captive of reptiles. The syndrome is commonly referred to as yellow fungus disease by bearded dragon breeders and fanciers. Lesions commonly start out as focal areas of crusting and scale loss and progress to more widespread deep, necrotic dermal lesions, that are commonly found on the face, inguinal region and digits (if present). Damage to the skin may eventually progress to infect the bone underneath, and thus can cause loss of digits, decreased appetite, or even decrease the animal's ability to apprehend food due to osteomyelitis in the jaw. These symptoms can lead to progressive wasting and potential dissemination of the organisms to the internal organs, as well as potential septicemia from secondary bacterial infection.

Some unusual aspects of this case highlight potential pitfalls when testing for and diagnosing yellow fungus disease: CASE 1 testing positive for *Adenovirus* while CASE 2 was negative despite being housed in the same enclosure and being purchased from the same location. Potential reasons for a negative test in this case are error in sample collection or shipping of the sample, lab error, and variance in sensitivity and specificity of the test, or a true negative.

It is also unusual that both dragons were cloacal PCR-negative for CANV while being confirmed histologically for CANV (although not speciated). A potential reason for the negative result could be that the PCR was run for a species of *Chrysosporium* that was not the species that the patient had, thus delivering the negative result.

Recent research has reclassified what was once labeled CANV (or the cause of yellow fungus disease into many species-specific isolates of *Nannizziopsis*; *Paranannizziopsis* and *Ophidiomyces*, the bearded dragon isolate being labeled *Nannizziopsis guarroi*,[2] although recent case reports have mentioned other species causing disease in bearded dragons.[3] The dragon's fungus in this report was not speciated, so it is not known which isolate caused the infection.

CASE 2 had a fungal-positive culture that was identified as a *Trichophyton*. *Chrysosporium* species can be difficult to identify from many other species of fungus as their conidia look similar to the microconidia produced by *Trichophyton* species. This may have caused the misidentification.[4]

This case also highlights the need to confirm a clinical cure with histopathology. It was shown that a patient can look clinically normal and still have low numbers of fungal organisms.

Side effects of voriconazole in humans include visual disturbances (hallucinations), gastrointestinal issues, hepatic toxicity, fetal toxicity in pregnant women, and electrocardiac disturbances.[5] Side effects seen in dogs include increased AST levels, liver enlargement, and significant impacts on other drugs that are metabolized by the liver. Side effects in cats include azotemia, anorexia, lethargy, weight loss, cutaneous drug reactions, hind limb paresis, ataxia, paraplegia, visual signs, arrhythmias, and hypokalemia.[6] Elevations in AST have been reported in bearded dragons, pigeons, mice, rats, dogs and African grey parrots on voriconazole.

It is unclear whether specific side effects can be attributed to the treatment with CASE 2 of this report, except that the patient did not like being medicated, became more reclusive, with decreased frequency of basking. The owner at the time attributed this to the bearded dragon's natural instincts to go into seasonal brumation as it was wintertime, and now has noticed a reversal of these behaviors as it is spring; however, we cannot rule out at this time that these observations were not side effect of the medication. There were no other observed clinical abnormalities.

Chrysosporium spp. is a large group of fungal pathogens of reptiles that can cause severe disease, and be clinically challenging and expensive to diagnose. Cheaper methods of diagnosis such as PCR may not be accurate, and may lead to a delay in proper diagnosis and initiation of treatment. Histopathology is currently the gold standard for diagnosis of this disease. Concurrent illnesses such as adenovirus may decrease prognosis, and diagnostic testing prior to beginning treatment is imperative to case selection.

References

1. Waeyeyenberghe L, Baert K, Pasmans F, Van Rooij P, Hellebuyck, Beernaert L, Backer P, Haesebrouck, Martel A. 2010. Voriconazole, a safe alternative for treating infections caused by the *Chrysosporium* anamorph of *Nannizziopsis vriesii* in bearded dragons (*Pogona vitticeps*). Med Mycol. 2010;28:880-885.

2. Sigler L, Hambleton S, Pare J. Molecular characterization of reptile pathogens currently known as members of the *Chrysosporium* anamorphs of *Nannizziopis vriesii* complex and relationship with some human-associated isolates. 2013. J Clin Microbiol. 51(10):3338-3357.

3. Schmidt-Ukaj S, Loncaric I, Spergser J, Richter B, Hochleithner M. Dermatomycosis in three central bearded dragons (*Pogona vitticeps*) associated with *Nannizziopsis chlamydospora. J Vet Diagn Invest.* 2016;28(3):319-322.

4. Cabanes FJ, Sutton DA, Guarro J. *Chrysosporium*-related fungi and reptiles: a fatal attraction. PLoS Pathog 10(10): e1004367.

5. Mader DR, Divers SJ, *Current Therapy in Reptile Medicine and Surgery*. St. Louis, MO: Elsevier Saunders;2014.

6. Plumb DC. *Plumb's Veterinary Drug Handbook*. 8th ed. Ames, IA: Wiley-Blackwell;2015.

Complete Edentulism in a Green Tree Python *(Morelia viridis)*

Bradley J. Waffa, MSPH, DVM,
Drury Reavill, DVM, Dipl ABVP (Avian), Dipl ABVP (Reptile and Amphibian), Dipl ACVP,
Stephen L. Barten, DVM,
Scott Echols, DVM, Dipl ABVP (Avian)

Session #141

Affiliation: From Churchland Animal Clinic, 6030 High Street W, Portsmouth, VA 23703, USA (Waffa), Zoo/Exotic Pathology Service, 6020 Rutland Drive #14, Carmichael, CA 95608, USA (Reavill), Vernon Hills Animal Hospital, 1260 Butterfield Rd, Mundelein, IL 60060, USA (Barten), and the Medical Center for Birds, 3805 Main Street, Oakley, CA 94561, USA (Echols).

A 5-year-old captive-bred female green tree python (*Morelia viridis*) was presented with a 2-week history of inappetence and concern for respiratory disease due to audible respiratory sounds. Aerobic culture/sensitivity of a tracheal wash sample revealed a *Pseudomonas* respiratory infection that was sensitive to and ultimately responded to treatment with ceftazidime. It also was discovered in the course of the work-up that the snake had lost its entire polyphydont dentition with no other apparent oral pathology, a phenomenon that to the best of the authors' knowledge has not been previously reported in the literature. High-resolution images and bone-density data generated using micro-CT were used on this case and compared with skulls from normal green tree python heads. All the heads and representative tissues were evaluated histologically.

Microspordial Keratoconjunctivitis in a Bearded Dragon *(Pogona vitticeps)*

Anna Martel-Arquette, DVM,
Sue Chen, DVM, Dipl ABVP (Avian),
Julie Hempstead, DVM, Dipl ACVO,
Rebecca Pacheco, DVM,
Natalie Antinoff, DVM, Dipl ABVP (Avian),
Leandro Teixeira DVM, MSc, Dipl ACVP

Session #200

Affiliation: Gulf Coast Veterinary Specialists, Houston, Texas, 77064, COPLOW - University of Wisconsin-Madison, Madison, WI 53706, USA.

Systemic microsporidiosis has been documented in bearded dragons.[1] However, to the authors' knowledge, there is no case of isolated, primary ocular protozoal infection in bearded dragons in the literature, nor any cases of successful treatment of microsporidiosis in reptiles.[1-2] A four-month old bearded dragon (*Pogona vitticeps*) presented with unilateral conjunctivitis that was not responsive to treatment with topical antibiotic therapy and oral and topical anti-inflammatories. Culture and sensitivity results indicated the presence of *Salmonella* species, which was sensitive to prescribed antibiotics. A biopsy of the conjunctival tissues was performed secondary to non-resolution of symptoms. Histopathology of the conjunctival tissue demonstrated Microsporidium (protozoal) organisms. Fenbendazole (50 mg/kg PO q24h for 5 days), systemic itraconazole (5 mg/kg PO q24h for 30 days), and topical itraconazole (1% ophthalmic ointment applied to left eye q24h until directed otherwise) was prescribed. The conjunctivitis continued to progress despite medical therapy and because of this, the eye was enucleated. Post enucleation, itraconazole was continued for an additional 3 months and fenbendazole was repeated (50 mg/kg PO q24h for 5 days, then q7days for 3 additional doses). Computed tomography was used to monitor tissues for potential central nervous tissue infection post enucleation. The patient remains asymptomatic 9 months post enucleation. This report documents the first successful treatment of microsporidiosis in a reptile. Microsporidiosis should be a differential diagnosis in keratoconjunctivitis cases that are not responsive to antibiotics.

References

1. Jacobson ER, Green DE, Undeen AH, Cranfield M, Vaughn KL. 1998. Systemic microsporidiosis in inland bearded dragons (*Pogona vitticeps*). *J Zoo Wildl Med.* 29(3):315-323.

2. Vergneau-Grosset C, Larrat S. 2015. Microsporidiosis in vertebrate companion exotic animals. *J Fungi.* 2(1):3-23.

Urinary Bladder Fistula via Plastronotomy Site in a Radiated Tortoise *(Astrochelys radiata)*

Daniel Cutler, DVM

Session 253

Affiliation: From University of Georgia, 2200 College Station Rd., Athens, GA 30602, USA.

Reproductive and urinary diseases are common presentations for captive chelonian species. Dystocia and urinary calculi are examples of such diseases that may develop from poor husbandry. Medical management may be possible for some cases, but surgical intervention often is unavoidable. The presence of the shell makes coelomic access difficult; however, there are multiple ways to access the reproductive and urinary organs. Endoscopic evaluation of the cloaca and urinary bladder via a cloacal approach is less traumatic, allows visualization of the bladder and therapeutic intervention in some situations. For larger calculi or egg removal, surgical approach either through the pre-femoral space or with a plastronotomy may be performed. The pre-femoral approach provides a limited amount of visualization and surgical access, compared to the plastronotomy, but is far less traumatic and requires shorter healing time.

This case of a radiated tortoise (*Astrochelys radiata*) provides examples of these techniques for a case of dystocia and complications. Use of both endoscopic and traditional surgical techniques used in conjunction can be useful in correction of egg binding. Both of these have risks and examples of such occurred in this case. Repeat endoscopy, after a pre-femoral correction of retro-pulsed eggs into the urinary bladder resulted in perforation of the bowel. Following a plastronotomy to address the perforation and perform an oophorectomy, a communication from urinary bladder to the plastronotomy site was diagnosed weeks later; however, in both cases they were identified and appropriately addressed, resulting in a positive outcome for the patient.

Subcarapacial Squamous Cell Carcinoma in a Hermann's Tortoise *(Testudo hermanni)*

Anke C. Stöhr, med. vet., ZB Reptilien,
Ulrich Freund, Dr. med., FA Radio-Onkologie,
Kim O. Heckers, Dr. med. vet.

Session #237

Affilation: Kleintierklinik Egelsbach, Boschring 8, 63329 Egelsbach, Germany and the Klinikum Offenburg, Weingartenstr. 70, 77654 Offenburg, Germany and Laboklin GmbH & Co. KG, Laboratory for Clinical Diagnostics, Steubenstr. 4, 97688 Bad Kissingen, Germany.

A female Hermann's tortoise was presented with a deep lesion in the caudal part of the carapace. Myiasis and bacterial infection were found and treated. Several weeks after the therapy, increased subcarapacial tissue growth was detected. No pathological changes were found in blood analysis. Tissue samples were taken under general anesthesia and submitted for bacteriological, mycological and pathohistologic examination. A squamous cell carcinoma was diagnosed. Tumor growth increased rapidly after sampling. A CT scan was performed to assess the complete tumor size and to rule out metastases. Percutaneous radiotherapy was performed for seven weeks using a linear accelerator (LINAC). Each week, five consecutive days of treatment were followed by two days without radiation. 60 Gy have been applied via photons of the LINAC (6 MV) homogenously in the area of the carcinoma including parts of the carapace. The therapy stopped tissue growth successfully and led to drying and mild shrinking of the tumor, but did not result in complete remission. The animal tolerated the therapy well. Several months later, the tortoise was euthanized due to severe myiasis. This is the first description of the use of radiotherapy in squamous cell carcinoma in a tortoise. Although the therapy did not cure the animal, the observed effects and the good tolerance indicated that this treatment approach is promising. In future cases, surgical removal of tumor tissue before radiation may optimize the therapeutic effect.

Cloacal Prolapse, Colonic Adenocarcinoma, and Hypercalcemia in a Ball Python *(Python regius)*

Danielle K. Tarbert, DVM,
Ricardo de Matos, LMV, MSc, Dipl ABVP (Avian), Dipl ECZM (Avian),
Dipl ECZM (Small Mammal),
Elizabeth Buckles, DVM, PhD, Dipl ACVP

Session #157

Affiliation: From the Department of Clinical Sciences, Section of Zoological Medicine (Tarbert, de Matos) and the Department of Biomedical Sciences, Section of Anatomical Pathology (Buckles), College of Veterinary Medicine, Cornell University, Ithaca, NY 14853, USA.

Abstract: An approximately 12-year-old male intact ball python (*Python regius*) was presented for a recurrent cloacal prolapse. The prolapsed tissue, determined to be colon, had significant necrosis and included a firm, thickened region. A resection and anastomosis of the colon was performed using an external approach. Ionized hypercalcemia was documented at the time of surgery. Histopathologic evaluation of the resected tissue revealed colonic adenocarcinoma. The snake recovered quickly and did well at home for at least 1 month. Colonic adenocarcinoma should be considered as a differential for cloacal prolapse in ball pythons, and evaluation for hypercalcemia of malignancy should be considered in ball pythons diagnosed with neoplasia.

Introduction

Cloacal prolapses in reptiles may involve the gastrointestinal tract, reproductive organs, or urinary bladder.[1] Identification of the prolapsed organ as well as the inciting cause is essential for successful treatment.[1] Prolapsed colonic tissue can be differentiated from the bladder, phallus (chelonians and crocodilians), or hemipenes (squamates) by the presence of a lumen.[1] A prolapsed oviduct will also contain a lumen; however, in most cases the colon can be identified by the presence of fecal material, either passed directly through the lumen or aspirated through a catheter.[1] Colonic prolapse is usually the result of tenesmus, with the most common causes including constipation, bacterial enteritis, and parasitic enteritis.[1] Constipation associated with colonic adenocarcinoma has been reported in a corn snake (*Elaphe guttata guttata*), although prolapse did not occur in that case.[2] This case report describes a colonic prolapse in a ball python (*Python regius*) secondary to colonic adenocarcinoma. Additional findings included hyperuricemia and hyperphosphatemia. Ionized hypercalcemia was present suggestive of hypercalcemia of malignancy. To the authors' knowledge, this is the first report of colonic adenocarcinoma in this species, and the first report of colonic neoplasia as the primary etiology in a cloacal prolapse. Antemortem diagnosis and treatment of the adenocarcinoma is discussed, as well as considerations for future research into the mechanisms of calcium homeostasis in reptiles.

Case Report

An approximately 12-year-old male intact ball python (*P regius*) was presented for evaluation of a cloacal prolapse which had occurred within the previous 5 days. The snake had a history of a previous prolapse 1 year before which had resolved with manual reduction by a local veterinarian. The owner also reported gradual weight loss over the past year. The snake had last eaten 3 weeks prior to presentation.

Physical examination revealed thin body condition and dehydration. A 5-cm diameter spherical mass was observed protruding from the vent. The proximal aspect of the mass was pink and edematous; the distal aspect was covered in a gray layer of fibrinonecrotic tissue. A lumen was found within the tissue; upon examination, feces were found within the lumen resulting in identification of the prolapsed tissue as colonic in origin. A sexing probe was passed into the hemipenial lumen to the level of the 10th subcaudal scales, confirming the snake was male.

Following administration of topical and systemic analgesics as well as antibiotics, the exposed tissue was lavaged and the necrotic cap debrided. Whole body radiographs were unremarkable. A fecal wet mount and direct smear showed no evidence of parasites. The snake was placed under general anesthesia for further evaluation of the prolapse. The dorsal aspect of the prolapsed colonic tissue was markedly thickened and firm. After further debridement of necrotic material, most of the prolapsed tissue was determined to be devitalized. An external resection and anastomosis was performed using previously described techniques.[1,3] The resected tissue was submitted for histopathology.

Cardiocentesis was performed prior to recovery from anesthesia. Results of an in-house plasma chemistry panel revealed hyperphosphatemia (8.3 mg/dl, reference interval 1.4-7.3 mg/dl) and hyperuricemia (12.2 mg/dl, reference interval 0.8-8.3 mg/dl).[4] Total calcium was 17.5 mg/dl (reference interval 10.4-19.3 mg/dl).[4] Results of an ionized calcium measurement were 2.09 mmol/L. Packed cell volume was measured at 14% (reference interval 10.4-33.4%) with a total solids of 5.4 g/dl (reference interval 3.6-9.0 g/dl).[4]

The python recovered well from anesthesia and was discharged the following day with prescriptions for subcutaneous fluids, anti-inflammatories, and antibiotics pending biopsy results. Histopathology revealed a transmural colonic adenocarcinoma in addition to necrosis of the mucosa with heterophilic colitis and fibrin thrombi. The neoplastic cells extended to the borders of the examined issues. The snake did well at home with good appetite and no further prolapses for at least one month before being lost to follow-up.

In the case described in this report, the identity of the prolapsed structure was readily determined based on the species, sex of the patient, and examination of the tissue. As all snakes lack a bladder, the tissue was immediately known to be either reproductive or colonic in origin.[1] Confirmation of male sex via probe eliminated prolapsed oviduct as the source of the tissue, while the presence of fecal material confirmed the structure to be colon. Two common causes of colonic prolapse, constipation and intestinal parasitism, were ruled down based on results of the radiographs and fecal analysis.[1] The unremarkable initial diagnostics and firm texture of the prolapsed tissue prompted submission for histopathology.

In acute cases of colonic prolapse, the tissue can be manually reduced and held in place with cloacal sutures until the underlying cause of the prolapse is resolved.[1] However, if the venous return from the tissue is compromised, the resulting edema may make the prolapse difficult to reduce.[1] A coeliotomy may be required in order to apply sufficient traction to the prolapse; with this method, a colopexy can be performed concurrently.[1] In cases where the prolapsed colon becomes devitalized, resection and anastomosis is required.[1] Although this can be performed via coeliotomy, an alternative external approach has been described.[3] This approach utilizes a stent (such as a plastic syringe case) to hold the lumen of the colon open while mattress sutures are placed circumferentially in healthy tissue.[3] The colonic tissue is then transected distal to the sutures, creating an inverting end-to-end intestinal anastomosis.[3] In this case, the duration of the prolapse (up to 5 days) in addition to the presence of neoplastic tissue likely contributed to the edema and necrosis. The external approach to resection and anastomosis allowed rapid surgical resolution without opening the coelomic cavity.

Colonic adenocarcinomas have been reported in several colubrids as well as in a Burmese python (*Python bivattatus*).[2,5,6] To the authors' knowledge, this is the first report of a colonic adenocarcinoma in a ball python.

In previous reports of snakes with colonic adenocarcinomas, no evidence of metastasis was found.[5-6] However, metastatic disease from other enteric adenocarcinomas has been reported, such as from an intestinal papillary adenocarcinoma in an emerald tree boa (*Corallus caninus*).[7] Although signs of metastasis were not seen on radiographs in the present case, metastatic lesions may be present in organs whose parenchyma is not well evaluated on plain film radiographs, such as within the liver, kidneys, or coelomic wall.[7] Additional diagnostic modalities such as ultrasonography, computed tomography, or endoscopy should be considered in monitoring for metastasis in snakes.

A recent study evaluating the effects of ultraviolet light on calcium metabolism in ball pythons generated an ionized calcium reference interval of 1.67-1.94 mmol/L.[8] The ionized calcium level of 2.09 mmol/L in the ball python described in this report was therefore considered elevated, with the total calcium at the high end of the reference interval. Although total calcium levels in reptiles are known to vary with physiologic status, ionized calcium levels should be consistent.[9] Furthermore, most elevations in total calcium are associated with vitellogenesis in females.[9] Differentials for hypercalcemia in a male snake include vitamin D toxicosis, renal disease, primary hyperparathyroidism, pseudohyperparathyroidism (hypercalcemia of malignancy), and osteolytic bone disease.[9] Vitamin D toxicosis and osteolysis were ruled down in this snake based on history and radiographs, respectively. Primary hyperparathyroidism was ruled down based on the concurrent hyperphosphatemia.[10] Renal disease may have been present in this case based on the hyperuricemia and hyperphosphatemia; however, in domestic species, renal disease is more commonly associated with low to normal ionized calcium values.[10] The elevated renal values may also have been associated with dehydration, post-renal obstruction from the prolapse, or injury from circulating calcium levels.[10] Although to the authors' knowledge hypercalcemia of malignancy has never been confirmed in a reptile, it was considered the most likely etiology of the hypercalcemia in this case.

Measurement of parathyroid hormone and parathyroid hormone-related protein can be utilized to definitively diagnose hypercalcemia of malignancy.[10] The previously available two-site immunoradiometric assay for measurement of parathyroid hormone was not useful in reptiles; however, new technology has recently become available.[10] Further studies are warranted to evaluate and, if possible, validate the use of commercially available parathyroid hormone assays in reptiles. This will allow more accurate diagnosis and monitoring of neoplastic and renal diseases as well as other diseases common in captive reptiles such as nutritional secondary hyperparathyroidism.

Disclosure Statement: The authors have no conflicts of interest to disclose.

References

1. Bennett RA, Mader DR. Cloacal prolapse. In: Mader DR, ed. *Reptile Medicine and Surgery*. 2nd ed. St. Louis, MO: Elsevier Saunders;2006:751-755.

2. Latimer KS, Rich GA. Colonic adenocarcinoma in a corn snake (*Elaphe guttata guttata*). *J Zoo Wildl Med.* 1998;29(3):344-346.

3. Leash AM. Amputation of a prolapsed rectum in an African rock python. *J Am Vet Med Assoc.* 1977;171(9):980-981.

4. International Species Inventory System. "Python_regius_No_selection_by_gender_All_ages_ combined_ Conventional_American_units_2013_CD.html" In: Teare, JA. *ISIS Physiological Reference Intervals for Captive Wildlife: A CD-ROM Resource.* Eagan, MN: International Species Information System;2013.

5. Sykes JM, Trupkiewicz JG. Reptile neoplasia at the Philadelphia Zoological Garden, 1901-2002. *J Zoo Wildl Med.* 2006;37(1):11-19.

6. Chandra AMS, Jacobson ER, Munn RJ. Retroviral particles in neoplasms of Burmese pythons (*Python molurus bivittatus*). *Vet Pathol.* 2001;38(5):561-564.

7. Orós J, Lorenzo H, Andrada M, Recuero J. Type A-like retroviral particles in a metastatic intestinal adenocarcinoma in an emerald tree boa (*Corallus caninus*). *Vet Pathol.* 2004;41(5):515-518.

8. Hedley J, Eatwell K. The effects of UV light on calcium metabolism in ball pythons (*Python regius*). *Vet Rec.* 2013;173(14):345-348.

9. Campbell TW. Clinical pathology. In: Mader DR, Divers SJ, ed. *Current Therapy in Reptile Medicine and Surgery*. St. Louis, MO: Elsevier Saunders;2014:70-92.

10. Schenck PA, Chew DJ, Nagode LA, Rosol TJ. In: DiBartola SP, ed. *Fluid, Electrolyte, and Acid-Base Disorders in Small Animal Practice.* St. Louis, MO: Elsevier Saunders;2012:120-194.

Section 30

ARAV Conservation

Anneliese Strunk, DVM, DABVP (Avian);
Tony Qureishi, DVM
Moderators

Follicle-Stimulating-Hormone-Induced Mating: Behavior and Histologic Changes in a Pair of Ocellated Lizards *(Timon lepidus)*

Emanuele Lubian, DVM, GPCert(ExAP),
Alessandro Vetere, DVM,
Massimo Millefanti, DVM

Session #005

Affiliation: From Ambulatorio veterinario, via Galvani 42, Gaggiano, 20083, Italy.

Abstract: This report describes a clinical approach and observed behavioral and histologic changes in a pair of *Timon lepidus* treated with follitropin alfa to induce breeding. Hormonal treatments are very rarely used to treat infertility in reptiles; otherwise in other species as well as in human medicine they, as artificial insemination, are one of the most important solutions. The conclusion of this case report suggests a possible therapy for this problem.

A pair of 3-year-old ocellated lizards (*Timon lepidus*) was evaluated for mating reluctance. After hibernation, the male was introduced into the female's cage. The female immediately showed aggresive behavior toward the male. This same procedure was successively tried without any positive results. After multiple failed attempts and about 1 month after the end of hibernation, we initiated therapy. We used follitropin alfa, a hormone identical to follicle-stimulating hormone (FSH), obtained by DNA recombination of ovarian cells of the Chinese hamster (*Cricetulus griseus*). Follitropin alfa often is used in human medicine in patients afflicted by hypogonadism or insufficient plasma levels of gonadotropins. By the eleventh day of treatment, the owner observed a behavior change that could be referred to as reproductive activity of the male. The female did not exhibit any aggressive behavior toward the male. Ten days after treatment, the owner reported the death of both specimens, due to a management mistake (i.e., both subjects were housed in the sun without shelter from excessive heat). Postmortem examination showed an increase in size of the gonads of both animals and no other organ alteration.

In conclusion, in this clinical case it is possible that the increase in testicular activity and sexual segment of the kidney, could be due to the administration of FSH. As a result we can assert that, at least in the organs taken into consideration, there has not been found any toxicity, linked to the administration of gonadotropin in the short period. This case report could be used as a starting point for additional studies of mating in captive reptiles.

References

1. Edwards TM, Gunderson MP, Milnes MR, Guillette, LJ. Gonadotropin-induced testosterone response in peripubertal male alligators. *Gen Comp Endocrinol*. 2004;175(2):251-258.

2. Khan UW, Rai U. Role of gonadotropin and Leydig cell-secreted factors in the control of testicular macrophage activities in the wall lizard, *Hemidactylus flaviviridis*. *Dev Comp Immunol*. 2008;32(4):348-355.

3. Khan UW, Rai U. In vitro effect of FSH and testosterone on Sertoli cell nursing function in wall lizard, *Hemidactylus flaviviridis* (Rüppell). *Gen Comp Endocrinol*. 2004;136(2):225-231.

4. Moore BC, Forouhar S, Kohno S, Botteri NL, Hamlin HJ, Guillette LJ. Gonadotropin-induced changes in oviducal mRNA expression levels of sex steroid hormone receptors and activin-related signaling factors in the alligator. *Gen Comp Endocrinol.* 2012;135(3):372-380.

5. Jones RE, Austin HB, Lopez KH, Rand MS, Summers CH. Gonadotropin-induced ovulation in a reptile (Anolis carolinensis): histological observation. *Gen Comp Endocrinol.* 1988;72(2):312-322.

6. Mitchell M, Funcke S, Bublat A, Fischer D, Lierz M. Determining the value of electroejaculation as a method of semen collection in lizards and chelonians. *Int Conf Avian Herpetol Exot Mamm Med.* 2015;426.

7. Olivieri M, Alvarez M, Anel L, Perez V, Caso M. Sperm extraction and cloacal anatomy of North American colubrids. *Int Conf Avian Herpetol Exot Mamm Med.* 2015;350.

Health Assessment Training Class for Desert Tortoise *(Gopherus agassizii)* Translocation Assessment and Monitoring

Jay Johnson, DVM,
Nadine Lamberski, DVM, Dipl ACZM,
Peregrine Wolff, DVM,
Kristina Drake, MS,
Kimberleigh Field, MS,
Roy Averill-Murray, MS

Session #323

Affiliation: From Arizona Exotic Animal Hospital, Mesa, AZ 85201, USA (Johnson), San Diego Zoo Safari Park, Escondido, CA 92027, USA (Lamberski), Nevada Department of Wildlife, Reno, NV 89511, USA (Wolff), US Geological Survey, Henderson, NV 89074, USA (Drake), US Fish and Wildlife Service Desert Tortoise Recovery Office, Reno, NV 89502, USA (Field, Averill-Murray).

The Mojave desert tortoise (*Gopherus agassizii*) is federally protected as threatened under the Endangered Species Act. Recently, numerous energy-development projects (primarily solar and wind) have gained approval to construct facilities in areas inhabited by desert tortoises. In an attempt to minimize the impact on desert tortoises, many projects propose to displace tortoises from the project sites and relocate them to other areas. Many steps must be taken to minimize the risks to the tortoises being moved, as well as to tortoises they might encounter after translocation from the project site. Among the many precautionary steps is assessing the health of tortoises at both the project and recipient sites, as well as at any reference sites as may be required for specific projects. A standardized approach to health assessment data collection is necessary in order to monitor populations across projects and over temporal and spatial scales. In efforts to train biologists to perform consistent health assessments, sample collection, data collection, and reporting, the Health Assessment Training Class was developed. Students are trained on proper tortoise handling, biosecurity, anatomy and diseases, identification of abnormal clinical signs, body condition scoring, subcarapacial venipuncture and other sample collection, sample processing, and overall data collection and reporting. Students are also trained on rehydration techniques, including epicoelomic fluid administration. There are both lectures and hands on training followed by a written and hands on practical exam. Since the first class in 2011, 114 students have completed the class and 68 currently are certified to perform the health assessments.

A First Epidemiologic Study on Main Pathogens of Wild Amphibians in Greece

P. Azmanis, DVM, Dr. med. vet, Dipl ECZM (Avian),
I. Strachinis, BSc (Biology),
P. Lymberakis, BSc (Biology), PhD (Taxonomy),
RE Marschang, Dr. med. vet, PD, Dipl ECZM (Herpetology)

Session #233

Affilliation: Dubai Falcon Hospital, Dubai, UAE (Azmanis), Biology Faculty, Aristotle´s University of Thessaloniki, Greece (Strachinis), National History Museum of Crete, University of Crete, Greece (Lymberakis), LABOKLIN GmbH, Bad Kissingen, Germany (Marschang).

The chytrid fungi (*B. dendrobatidis, B. salamandrivorans*) and viruses (*Ranavirus* sp) are responsible for mass mortalities in amphibians.[1,2] Recently many field and experimental studies in West Europe have shed light into the epidemiology and pathogenesis of these microorganisms. So far, similar studies and data are scarce[3,4] or absent from East Europe and the Balkan Peninsula. Greece, in the southern tip of the Balkan Peninsula, offers a unique habitat for 23 amphibian species. Few of them (*P.cretensis, P. ceringensis, L. helverseni*) are endemic and their population is considered endangered or fragile.[5] Our team has sampled more than 300 amphibians of different species and habitats, from mainland and insular Greece, in order to investigate the possible presence and distribution of these pathogens. From each individual, a skin, oral and cloacal swab was collected under standard methodology. Endemic and endangered animals received special attention in our study. During the field sampling the team followed international biosecurity guidelines. All amphibians were returned unharmed to their original habitat. The samples were examined with conventional and real-time PCR, following standard methodology.[6,7] So far we have traced *B. dendrobatidis* in 7% of the examined anurans (*Bufotes viridis*, *Pelophylax ridibundus*, *P. epeiroticus*, *P. cretensis* and *Rana catesbeiana*) from three different areas. No other pathogens were so far detected. The future of the project and the latest results will be presented and discussed.

References

1. Murray KA, Retallick RWR, Puschendorf R, Skerratt LF, Rosauer D, et al. Assessing spatial patterns of disease risk to biodiversity: implications for the management of the amphibian pathogen, *Batrachochytrium dendrobatidis*. *J Anim Ecol*. 2011;48:163-173.

2. Olson DH, Aanensen DM, Ronnenberg KL, Powell CI, Walker SF, et al. Mapping the global emergence of *Batrachochytrium dendrobatidis*, the amphibian chytrid fungus. *PLoS One* 2013;8(2):e56802.

3. Vörös J, Bosch J, Hartel T. First record of *Batrachochytrium dendrobatidis* on amphibians in Romania. *J Zool*. 2013;9:446-449.

4. Vörös J, Jelic D. First steps to survey chytrid fungus in Croatia. *HYLA*. 2011;1:31-34.

5. Valakos E, Pafilis P, Sotiropoulos K, Lymberakis P, Maragou P, Foufopoulos J. *The Amphibians and Reptiles of Greece*. Frankfurt, Germany: Chimaira Editions;2008.

6. Blooi M, Pasmans F, Longcore JE, Spitzen-van der Sluijs A, Vercammen F, Martel A. Duplex real-time PCR for rapid simultaneous detection of *Batrachochytrium dendrobatidis* and *Batrachochytrium salamandrivorans* in amphibian samples. *J Clin Microbiol*. 2013;51(12):4173-4177.

7. Boyle DG, Boyle DB, Olsen V, Morgan JA, Hyatt AD. Rapid quantitative detection of chytridiomycosis (*Batrachochytrium dendrobatidis*) in amphibian samples using real-time Taqman PCR assay. *Dis Aquat Organ*. 2004;60(2):141-148.

Evaluating Thyroid Function and Establishing Reference Intervals in Endemic *Galápagos* Giant Tortoises

Adrien M. Zap, MS, DVM,
Isabelle L. Desprez, DVM,
F. David Bravo Rada, MVZ,
Andrés G. Ortega Ojeda, MBA, DMVZ,
Ellen N. Behrend, VMD, MS, PhD, Dipl ACVIM,
Sylvain Larrat, DVM, DES, Dipl ACZM,
E. Marie Rush, DVM, Dipl ACZM

Session #055

Affiliation: From Department of Pathobiology, St. George's University, School of Veterinary Medicine, True Blue, St. George's, Grenada, West Indies (Zap, Rush), Centre Hospitalier Universitaire Vétérinaire d'Alfort, École nationale *vétérinaire d'Alfort*, 7 avenue du Général de Gaulle, 94704 Maisons Alfort, France (Desprez), Animalopolis Hospital Clínica Veterinaria, Laureles 203, Guayaquil, Ecuador (Bravo), Hospital Docente de Especialidades Veterinarias, Universidad San Francisco de Quito, Diego de Robles y Vía Interoceánica, *Cumbayá,* Ecuador (Ortega), Department of Clinical Sciences, Auburn University, College of Veterinary Medicine, 1500 Wire Road, Auburn, AL 36849 USA (Behrend), Clinique Vétérinaire Benjamin Franklin - 38 Rue du Danemark ZA, Porte Océane 2 56400 Brech/Auray, France (Larrat).

The endangered *Galápagos* giant tortoise (*Chelonoidis* species) is the largest extant tortoise species and a keystone member of its fragile island ecosystem.[1] Previous research has focused on genetics,[2] ecology,[3] or health of non-endemic tortoises,[4,5] yet basic reference information of these animals remains unknown. It is anecdotally accepted that *Galápagos* tortoises are predisposed to thyroid disorders, but to date, all endocrine studies in this species have been performed on small groups of captive born tortoises in the United States, under conditions dramatically different from their native island environment.[4-6] This study establishes reference intervals and evaluates thyroid hormone function in endemic *Galápagos* giant tortoises. Samples were collected from 131 endemic adult *Galápagos* tortoises, representing wild and contained populations of three species (*Chelonoidis chathemensis, C. guntheri, C. vicinia*) from two islands during July and August 2015. Each tortoise was examined, measured, and photographed prior to sample collection. Packed cell volume and total solids were manually assessed, and blood chemistry was analyzed with Abaxis VetScan VS2 and i-STAT point of care tests (Abaxis Inc, Union City, CA, USA). Secondary analysis (complete blood counts, parasite assessment and thyroid hormone radioimmunoassay) is underway at diagnostic laboratories, and results will be available by end of 2016. Statistical analysis performed with Software R (R Foundation, Vienna, Austria) determined 95% reference intervals for data, producing the first database of health parameters for endemic *Galápagos* tortoises. This allows for species-specific diagnosis, assessment and treatment of this unique tortoise.

References

1. Boyer TH, Boyer DM. Turtles, tortoises and terrapins. In: Mader DR, ed. *Reptile Medicine and Surgery*. 2nd ed. St. Louis, MO: Elsevier Saunders;2006:78-99.

2. Caccone A, Gibbs JP, et al. Origin and evolutionary relationships of giant *Galápagos tortoises. Proc Natl Acad Sci USA. 1999;96(23):13223-13228.*

3. Blake S, Yackulic CB, et al. Vegetation dynamics drive segregation by body size in Galapagos tortoises migrating across altitudinal gradients. *J Anim Ecol*. 2013;82:310-321.

4. *Norton TM, Jacobson ER, et al. Medical management of a Galapagos tortoises (Geochelone elephantopus) with hypothyroidism. J Zoo Wildl Med. 1989;20(2):212-216.*

5. *Fleming GJ, Heard DJ, et al. Thymic hyperplasia in subadult Galapagos tortoises, Geochelone nigra. J Herpetol Med Surg. 2004;14(1):24-27.*

6. Rostal DC, Robeck TR, et al. Seasonal reproductive cycle of the Galapagos tortoise (*Geocheleone nigra*) in captivity. *Zoo Biol*. 1998;17:505-517.

Section 31

ARAV Posters

Testudo hermanni: Breeding Techniques and Performances in Two Farms of Central Italy

G. Deli, DMV,
E. Lubian, DVM,
D. Petrini, DVM, GPCert (ExAP),
C. Manetti,
O. Olivieri

Session #118

Affiliation: From Rome, Italy, veterinary practitioner (Deli), Vigevano, Italy, veterinary practitioner (Lubian), Pisa, Italy, veterinary practitioner (Petrini), Firenze, Italy, veterinary practitioner (Manetti), Department of Veterinary Medicine, Perugia, Italy (Olivieri).

This study on *Testudo*'s feeding was conducted from May 2011 to September 2012. This wide period of observation allowed us to investigate aspects concerning not only management factors (such as diet) but also some external factors such as macro and micro-climate changes.

Because of their importance to our country this research involved only *Testudo hermanni hermanni* belonging to different age groups, situated in two farms (hereafter referred to as A and B). Animals were divided into different groups according to their year of birth (2008 to 2012). For each group we chose to observe 10 subjects recognizable through specific markings. Systematically before hibernation and every time the animals woke up we gathered morphometric data (weight and size). The farmers cooperated and provided information concerning the organization, births, mortality, surgeries and pathologies. As each farm worked with different management systems we carried out a different protocol for each situation.

Farm A was handled by an amateur breeder who recorded type and amount of food administered to animals on a daily basis. Farm B maintained a greater number of turtles, as it is an extensive type of breed, so we carried out a less thorough and systematic data collection. In both farms turtles were fed a natural diet.

Striking differences are certainly due to a number of contingent factors such as the type of animal management, the different treatments, feeding, the environmental differences faced by animals and, above all, the genetics of the various groups.

References

1. Avanzi M. *Guida alle Tartarughe terrestri, le Guide Pet*. Milan, Italy: De Vecchi;2002:6-18.

2. Avanzi M. *Tartarughe terrestri*. Milan, Italy: De Vecchi;2011:19-63,71-83,112-113.

3. O'Malley B. *Clinical Anatomy and Physiology of Exotic Species*. St. Louis, MO: Elsevier Saunders;2005:18-35,41-48,92,453.

4. Girling SJ, Raiti P. *BSAVA Manual of Reptiles*. 2nd ed. Quedgeley, Gloucester: Br Small Anim Vet Assoc.;2004:75.

5. Mader DR. Reptile Medicine and Surgery. 2nd ed. St. Louis, MO: Elsevier Sauders;2006:704-764.

6. Nijboer J, Hatt J-M, Kaumanns W, Beijnen A, Ganslober U, eds. Zoo Animal Nutrition. Furth, Germany: Filander Verlag;2011:34-39.

7. Santoni L, Cavaterra M. *Testudo hermanni* in umbria. *Tartoombria*. 2007.

8. Ackermann L. *The Biology, Husbandry and Health Care of Reptiles*. Somerset, NJ: TFH Publications;1998.

Adenovirus Infection Management in a Breeding Bearded Dragon Group: Diagnosis and Treatment

Emanuele Lubian, DVM GPCert(ExAP),
Alessandro Vetere, DVM,
Diana Binanti, DVM, PhD, Dipl ECVP,
Gianluca Deli, DVM, GPCert(ExAP),
Massimo Millefanti, DVM

Session #162

Affiliation: From Ambulatorio veterinario, Via Galvani 42, Gaggiano, 20083, Italy (Lubian, Vetere, Millefanti), From AbLab veterinary diagnostic laboratory, Via Privata Massa Neri 13, Sarzana 19038, Italy (Binanti), From Ospedale veterinario Gregorio VII, Piazza Di Villa Carpegna 52, Roma 00165, Italy (Deli).

Abstract: This case report starts with several deaths in young Pogona vitticeps. The goal of the clinical procedures was to understand which animals were infected within the breeding group and to take measures in order to avoid further infections. Diagnostic procedures carried out were: postmortem, visceral and cloacal swab, endoscopic biopsy from internal organs. Study conclusions show how different tests were not enough to diagnose adenoviral infection in bearded dragon.

This case report shows how the sensitivity of different diagnostic methods for Adenovirus infection in bearded dragons is very low, which means that it should be very difficult to diagnose it and, in breeding farms, to prevent the entrance of the virus into the breeding group. The clinical case began with several deaths in young *Pogona* vitticeps, aged about 2-3 months. The breeder wanted to understand, as soon as possible, the cause of the deaths to avoid diffusion of the disease. All adults were separately bred except for 3 breeding females, some subadults and young subjects born in that year (before the deaths). We decided to isolate all young lizards from each other, not just the nonhealthy subjects from the healthy animals, as the owner had done before. After that, we sent the dead animals to the lab for postmortem evaluation and we tested the sickest of the group for Adenovirus, Ranavirus and Reovirus.

The results of the swabs were positive for Adenovirus and negative for the other diseases. Post mortem showed no evidence for macroscopic alterations (except poor nutritional condition), microscopic examination showed a lot of intranuclear inclusion bodies in liver, esophagus, stomach and intestine with a concomitant catarrhal gastroenteritis characterized by infiltration of inflammatory cells (lymphocytes, macrophages and heterophils).

Based on these results, we were sure that the cause of the death was an adenoviral infection. Then, the goal of next procedures was to treat the two remaining young bearded dragons and to understand which were the carriers in the adult group.

Treatment of the young lizards included fluid therapy, antibiotics, and force feeding. Of the two young bearded dragons infected, just one survived and the postmortem of the other one was similar to the ones explained before and exhaustive of Adenovirus.

We chose 9 of 20 adults and sub-adults (closest to the infected). In this group there were also the females and the father of the young lizards (morph het witblits het translucent and leatherback het witblits het translucent). We tested them for the same viruses of the infected subjects and all 9 bearded dragons were negative for all viruses. The parents were tested again after 2 months but they were negative again.

After that, we decided to take biopsies from parents of dead animals for histologic analysis. The goal was to see if parents showed the same lesions and inclusion bodies as the dead bearded dragons, so we could be sure that they were carriers. We chose to take liver biopsies using an endoscope, as well as from the intestine and stomach because these organs were rich with inclusion bodies in the dead animals.

Animals were premedicated 8 hours before with tramadole 7 mg/kg IM and induced before endoscopy with alfaxolone 7 mg/kg EV and then mantained with isoflurane 2%. We started with the celioscopy to take samples from liver. The incision was made in the hypochondrium area a bit lateral from the ventral medial line. The coelomatic cavity was explored to look for any other alterations, and then 3 samples were taken from the liver using biopsy forceps. After that, other 3 samples were taken for every apparatus through gastroscopy and cloacoscopy. For all endoscopic exploration, a rigid endoscope (1,8-mm, 30° optical inclination) was used.

All biopsies did not show any intra-nuclear inclusion body, but an infiltration of the wall of inflammatory cells (macrophages, lymphocytes, rare heterophils) was evident in both animals, in particular in stomach and intestine. There were edema and fibrosis, too. In liver biopsies there was infiltration of inflammatory cells, as seen in other apparatus.

With those results, we could not identify the mode of transmission because we did not find any positive adult. Also, the young lizard that survived became negative for Adenovirus, as it was tested with a cloacal swab 2 months after the last death.

At the end, we decided to test animals after the brumation season, during which immunity defenses are usually less than normal. We took samples from cloacal swabs from parents and other adults or sub-adults closer to the infected subjects the year before. All of the results were negative. At same time, we took samples from the eggs of the female (mother of the infected subject from the previous year) to identify with PCR any vertical transmission. No positivity for Adenovirus was detected.

In summary, we can say that the goal of the clinical procedures was to understand which animals were infected within the breeding group, because there were really few possibilities that adenoviral infection came from outside (the breeder was very careful, in order to avoid diseases, in the management of the animals). The research to ascertain the carrier was difficult and after all clinical procedures we have not been able to identify it.

We can state for sure that Adenovirus is detectable in infected animals that are carriers, eliminators and spreaders of the virus (we found positivity both with cloacal swabs for PCR and in postmortem with the visualization of intranuclear inclusion bodies in different organs). We can also conclude that positive animals can become negative for cloacal PCR test in few months. In fact, in our case, the surviving young lizard was negative after 2 months.

At the end, contrary to other studies and reports (that isolate Adenovirus from healthy animals through cloacal swab, meaning that they are carriers and eliminators) in our experience it was not so easy to find infected animals that are clinically healthy. In this clinical case we were not able to identify any positivity (but very probably an animal was a carrier) with all diagnostic procedures we used.

References

1. Latney LV, Wellehan J. Selected emerging infectious diseases of squamata. *Vet Clin North Am Exot Animal Pract*. 2013;16:319-338.

2. Ball I, Hoferer M, Marschang RE. Establishment of agamid cell line and isolation of adenoviruses from central bearded dragon (*Pogona vitticeps*). *J Vet Diagn Invest*. 2014;26(2):221-225.

3. Grafl B, Aigner F, Liebhart D, Marek A, Prokofieva I, Bachmeier J, Hess M. Vertical transmission and clinical signs in broiler breeders and broilers experiencing adenoviral gizzard erosion. *Avian Pathol*. 2012;41(6):599-604.

4. Schillinger L, Mentre V, Nicolier A, Blahak S. Adenoviral infection in juvenile inland bearded dragons (Pogona vitticeps) coinfected with microsporidiosis (*Microsporidium* sp.) and intestinal coccidiosis (*Isospora amphiboluri*). *Proc Int Conf Avian Herpetol Exot Mamm Med*. 2015; 344.

5. Marschang RE. Viruses infecting reptiles. *Viruses*. 2011;3(11):2087-2126.

6. Moormann S, Seehusen F, Reckling D, Kilwinski J, Puff C, Elhensheri M, Wohlsein P, Peters M. Systemic adenovirus infection in bearded dragons (*Pogona vitticeps*): histological, ultrastructural and molecular findings. *J Comp Pathol*. 2009;141:78-83.

Salmonella enterica Subspecies *arizonae* and *salamae* Associated Osteomyelitis in a Collection of Corn Snakes *(Pantherophis guttatus)*

Tom Hellebuyck, DVM, PhD, Dipl ECZM (Herpetology),
Frank Pasmans, DVM, PhD, Dipl ECZM (Herpetology),
Francis Vercammen, DVM, PhD,
Marc Verlinden, DVM, PhD,
Annemie Van Caelenberg, DVM, PhD,
Caroline Bleacher,
Richard Ducatelle, DVM, PhD, Dipl ECVP,
An Martel, DVM, PhD, Dipl ECZM (Wildlife Population Health)

Session #178

Affiliation: From the Department of Pathology, Bacteriology and Avian Diseases (Hellebuyck, Pasmans, Verlinden, Bleacher, Ducatelle, Martel) and the Department of Medical Imaging (Van Caelenberg), Faculty of Veterinary Medicine, Ghent University, Salisburylaan 133, B-9820 Merelbeke, Belgium and Centre for Research and Conservation, Royal Zoological Society of Antwerp, Koningin Astridplein 26, B-2018 Antwerp, Belgium (Vercammen).

Based on radiographic examination, progressive debilitating osteomyelitis was diagnosed in two corn snakes (*Pantherophis guttatus*) with multiple subcutaneous bony nodules located at the dorsal midline and paresis. Four corn snakes that were housed in the same enclosure as the two affected snakes remained clinically healthy during a 6-month observation period. In the first corn snake *Salmonella enterica* subspecies *arizonae* serotype 44:Z4Z23 was isolated antemortem following microbiologic examination of biopsies of the affected vertebrae and necropsy revealed the presence of granulomatous osteomyelitis causing lumbal stenosis, hepatitis and orchitis with isolation of serotype 44:Z4,Z23. *Salmonella enterica* subspecies *salamae* serotype 18:Z4,Z23 was isolated from the vertebrae following necropsy of the second snake. Two months later a corn snake from a second group, comprising 6 corn snakes that were reared separately from the first group, died after a 2-week period of weight loss and anorexia. Necropsy revealed osteomyelitis of several vertebrae and *Salmonella enterica* subspecies *arizonae* serotype 44:Z4Z23 was cultured from the vertebral lesions and the liver. *Salmonella* could not be isolated from intestinal samples in the affected snakes or in fecal samples of the clinically healthy snakes from both groups. Both *Salmonella* serotypes appear to have a tropism for bone and other extraintestinal sites in corn snakes and may cause a progressive, ultimately fatal disease in this species.

Limb Amputation in a Crested Gecko (*Correlophus ciliatus*)

Emanuele Lubian, DVM, GPCert(ExAP),
Elena Ferlini, DVM,
Massimo Millefanti, DVM

Session #202

Affiliation: From ambulatorio veterinario, Via Galvani 42, Gaggiano, 20083, Italy.

Abstract: Limb amputation is a common solution adopted (and usually well tolerated) in several animals with different diseases. This procedure in geckos (which use limbs to climb on vertical walls) must consider possible behavioral repercussions and health after surgery. There are few reports that indicate when to perform the amputation, the possible surgical procedures and the quality of life after surgery. In this clinical case, a hind limb was amputated in a *Correlophus ciliatus* because it was necrotic. After this procedure, the gecko started to move, climb, and hunt as it did before surgery.

Limb amputation is a common solution adopted in different animals with a variety of diseases. It is always a drastic solution for the animal because of the loss of the entire limb, but this is usually well tolerated in many domestic animals. This procedure in a gecko that spends its life climbing on vertical walls must consider the possible ethologic repercussions and health after surgery.

An adult female of *C. ciliatus* was examined because it was trapped in its enclosure system in its terrarium for 5 days. After the trauma, the limb became necrotic. The gecko was anesthetized using mask induction with isoflorane 5% and maintained with isoflorane 3%. After that, it was cleaned with diluted chlorhexidine solution from the abdomen to the top of the tail and placed ventro-dorsally. The surgical procedure was similar to those used with other pets, but the 2 skin flaps were closed on the dorsal surface of the limb to avoid problems with healing caused by rubbing on the ground. After surgery, the gecko did not show any problems in movement, searching and attacking prey, both on vertical and horizontal walls. Sutures were removed after 8 weeks. After 1 year the animal is absolutely clinically normal.

References

1. Suedmeyer WK. Limb amputation in a tokay gecko (*Gekko gecko*). *Exot Pet Pract.* 1998;3(7):55.

2. Mader DR, Bennett AR, Funk RS, Fitzgerald KT, Vera R, Hernandez-Divers SJ. Limb amputation. In: *Reptile Medicine and Surgery*. 2nd ed. Philadelphia, PA: Elsevier Saunders;2006:609-610.

3. Autumn K, Liang YA, Hsieh ST, Zesch W, Chan WP, Kenny TW, Fearing R, Full RJ. Adhesive force of a single gecko foot-hair. *Nature*. 2000;405:681-685.

4. Clabaugh K, Haag M, Hanley CS, Latimer KS, Hernandez-Divers SJ. Undifferentiated sarcoma resolved by forelimb amputation and prosthesis in a radiated tortoise (*Geochelone radiata*). *J Zoo Wildl Med.* 2005;36(1):117-120.

5. Fossum TW, Hedlund CS, Hulse DA, Johnson AL, Seim HB, Willard MD, Carroll GL. Chirurgia ortopedica, In: *Chirurgia dei Piccoli Animali*. seconda edizione. Masson Spa Milano: EV srl Cremona;2004:1169-1170.

Section 32

Saturday Hybrid Master Classes

Troubleshooting Poor Anesthetic Recovery: Persistence Pays

Angela M. Lennox, DVM, Dipl ABVP (Avian), Dipl ABVP (Exotic Companion Mammal), Dipl ECZM (Small Mammal)

Session #257

Affiliation: From Avian & Exotic Animal Clinic, 9330 Waldemar Road, Indianapolis, IN 46268, USA.

Exotic companion mammals have a reputation for increased surgical risk. A 2006 study comparing anesthetic death rates of various pet species reported a rate of 0.17% for dogs, 0.24% for cats, 1.39% for rabbits, and 3.8% for guinea pigs.[1] Explanations for increased surgical death rate abound, and include greater intolerance to pain and stress, fewer numbers of well, elective procedures in this species, and the ability to hide underlying illness.

Surveys of traditional domestic pets indicates the majority of anesthetic deaths occur during the post-operative period. Careful monitoring and attention to detail can help rapidly identify the patient experiencing a less than ideal recovery; quick identification of underlying etiologies can lead to improved survival rates. This presentation emphasizes the role of the veterinary technician as part of a team-based approach to reducing loss during the recovery period.

Incidence of poor anesthetic recovery can be dramatically reduced by paying careful attention to pre- and peri-surgical details, which include the following:

1. Adequate patient preparation: Few surgeries in exotic companion mammals are urgent, with the exception of conditions such as urinary or gastrointestinal obstruction. In most cases there is time to perform a complete pre-anesthetic evaluation, including an examination just prior to surgery, as condition can rapidly change, and pre-anesthetic blood work.

2. Vascular access: Vascular access is considered a standard of care for human and traditional pet species undergoing any anesthetic procedure. With recent increased utilization of techniques such as placement of intraosseous catheters, vascular access should be attempted in every exotic surgical patient as well. A general recommended surgical fluid rate is 10 ml/kg/hr; however, ideal rates for each species have not been determined.

3. Careful selection of pre-anesthetic protocols: There is no legitimate indication for single agent inhalant anesthesia in exotic companion mammals. Pre-anesthetic agents have been shown to reduce the concentration of inhalant general anesthetics, which is linked with greater patient safety. Pre-anesthetics contain analgesics, which provide pre-emptive analgesia.

4. Careful selection of induction and maintenance agents.

5. Intubation of patients whenever possible.

6. Proficient and continuous anesthetic monitoring.

7. Adequate analgesia: As pain is associated with poor anesthetic recoveries in many species; multi-model analgesia should be considered, including pre-emptive analgesia, local blocks, constant rate infusion (CRI) of analgesics, and effective post-surgical administration.

Recovery

Patients should be monitored carefully post surgery, until they are up and about, walking, and taking food. Ideally, a designated recovery technician monitors recovering patients continuously at first, and then intermittently as recovery progresses. When recovery is delayed, every effort must be made to attempt to find a cause and provide a remedy. In the author's opinion, many patients are lost in the recovery period due to inadequate analgesia and hypovolemia.

Causes of Poor Recoveries

There are a number of factors associated with poor anesthetic recovery. These factors are described below.

Use of pre-anesthetic agents

Use of pre-anesthetic drugs usually results in a delayed recovery, when compared to no or fewer agents. This is often concerning in practices just beginning to add new protocols and drug combinations. Ideally, the recovery period is slow, steady and calm, as opposed to a rapid recovery with struggling. This must be distinguished from a poor recovery that could benefit from reversal of one or more pre-anesthetic agents; experience and familiarity with the protocol in a particular species are often the best guides. Agents commonly used in ECM that are reversible include dexmedetomidine and midazolam. The reversal of opioid drugs is a controversial; complete reversal of analgesic agents is never recommended; however partial reversal may be considered.

Hypovolemia

The hypovolemic patient demonstrates a low or unobtainable blood pressure. In patients where acquisition of blood pressure is impractical, other clinical markers should be considered, including decreased mentation, pale mucus membrane color and delayed capillary refill time. The hypovolemic patient is weak and responds poorly to stimulus. Optimal fluid therapy is critical for treatment of hypovolemia associated with anesthesia or blood loss. While little information exists on specific guidelines for treatment of hypovolemia in various exotic species, information can be extrapolated from with other traditional pet species. The endpoint of fluid resuscitation is restoration of normal blood pressure and normalization of other parameters. The following strategy outlines treatment of hypovolemia in exotic companion mammals

Reverse pre-anesthetic agents, if applicable (see above). Administer rapid intravenous infusion of warmed isotonic crystalloids at 10-15 ml/kg, followed by colloids (Hetastarch, 6%, B Braun Medical Inc., Irvine, CA) at 5 ml/kg over 5-10 minutes. Boluses of isotonic crystalloids (10-15 mg/kg) and colloids (5 ml/kg) are continued until systolic Doppler blood pressure reads above 60-70 mmHg.

An alternative to initial treatment with crystalloids is the use of 7.2-7.5% hypertonic saline at 3 ml/kg as a slow bolus. Hypertonic saline quickly draws fluids from other compartments into the intravascular space.[2] The effect is maintained with follow up administration of crystalloids as discussed above.

If blood pressure cannot be obtained, look for improvement in other clinical markers.

Anemia due to hemorrhage

The patient with ongoing blood loss post surgery presents similarly to the patient with hypovolemia; features include, pale mucus membranes, poor mentation and delayed CRT. The surgeon should have some indication that hemorrhage was possible based on the nature of the surgery or the recovery technician may observe ongoing blood loss from a surgery site.

Determination of PCV is usually not helpful in detecting acute blood loss. General fluid support may be adequate for treatment of mild to moderate anemia (see above).

More severe hemorrhage may require blood transfusion or the use of colloids with oxygen carrying ability such as Oxyglobin, which is currently unavailable in the United States. With the exception of the ferret, exotic companion mammals are reported to have distinct blood types; however, the likelihood of transfusion reaction after a single transfusion is not high. The risk of reaction must be weighed against the risk of withholding transfusion. Sources of blood donors include the ill pet's housemates, rescue organizations, or pet stores. The author keeps a list of owners willing to provide blood donors in exchange for clinic credit. Blood is collected from healthy donors under sedation if necessary with 1 ml acid citrate dextrose (ACD) per 10 ml blood, maximum 10% of blood volume based on calculated body weight. Whole blood is administered via IV or IO catheter.

An example of blood transfusion following hemorrhage is outlined below:

- Place an IV catheter or IO catheter of the recipient if not already established.
- Sedate or anesthetize the donor.
- Collect blood from the donor (7-10% of body weight) via the largest available and safely accessed vein into one or several 1-3 ml syringes prepared ACD 1cc/10 ml blood.
- Administer whole blood via the IO catheter manually over 3-5 minutes, or with a small precision infusion pump.

Hypoglycemia

True hypoglycemia is not commonly seen in rabbits or rodents, but may be seen post operatively in the ferrets with insulinoma. Even ferrets that demonstrate normoglycemia pre-surgically may experience marked hypoglycemia due to fasting prior to surgery or decompensation due to stress, anesthesia and surgery. It should be kept in mind ferrets undergoing pancreatic nodulectomy or partial pancreatectomy may also demonstrate marked hyperglycemia, which often takes 24-72 hours for normalize. Other causes of hypoglycemia include septicemia, which can occur in any patient, including rabbits and rodents.

Hypothermia

Loss of heat prior to and during surgery may continue well into recovery; therefore temperature should be monitored until the patient is upright and moving. Hypovolemic mammals below a certain temperature threshold are not able to obtain normalvolemia, and fluid resuscitation will not be successful. Rewarming methods include external warming and infusion of warmed IV fluids.

Some practitioners have noted hyperthermia in ferrets that may occur up to an hour or more post surgery. This may be similar to hyperthermia as a result of opioid administration in cats.

Pain and discomfort

Pain has long been associated with poor surgical outcomes in traditional pet species, and likely applies to exotic species as well. The painful patient differs from the hypovolemic patient in that the animal is generally not weak, but demonstrating tense, painful body posture. It should be kept in mind that the duration of analgesia of analgesics are unknown for all exotic species under all conditions; therefore, if analgesia is administered as part of pre-anesthesia, and the preparation/surgery period is more than a few hours, the patient may be recovering without benefit of analgesia. Multi-model protocols are extremely important (constant rate infusion (CRI) of analgesics, local analgesia). For patients with poor recoveries due to suspected pain, administer a dosage of an opioid IV, if possible, and monitor carefully for the next 15-30 minutes.

Cardiac disease

Animals with pre-existing occult cardiac disease may decompensate as a results of anesthesia and surgery. Patients in heart failure often exhibit respiratory distress. Careful handling and additional sedation may help with obtaining diagnostic imaging in order to confirm diagnosis.

Trouble Shooting

Treatment often is required for surgical patients experiencing slow recovery from anesthesia. A trouble-shooting guide for slow recoveries as a result of the factors described above is presented in Table 1.

Table 1. Trouble shooting guide for slow recoveries in surgical patients.

Potential cause	Diagnosis	Treatment
Administration of pre-anesthetics	Rule out other causes of slow recovery	None required; monitor carefully Reverse if prolonged
Pain	Rule out other causes of slow recovery; note painful posture with rapid cardiac and/or respiratory rate. Consider especially if time of last dose of opioid is more than 3-4 hours prior.	Administer low dose opioid IV or IO and observe response over next 10 minutes
Hypovolemia	Take indirect blood pressure if possible, observe peripheral vein turgor, evaluate mm/CRT	Administer crystalloids and colloids at shock rates IV or IO as indicated above
Hypoglycemia	Measure blood glucose with hand-held glucometer	Administer glucose PO if patient is swallowing; otherwise IV, IO
Hemorrhage	Look for evidence of hemorrhage, see section on hypovolemia above, note mucus membrane color	Administer colloids and/or whole blood

References

1. Brodbelt DC, Blissitt KJ, Hammond RA, Neath PJ, Young LE, Pfeiffer DU, Wood JL. The risk of death: confidential enquiry into perioperative small animal fatalities. *Vet Anaesth Analg*. 2008;35(5):365-373.

2. Bowles H, Lichtenberger M, Lennox A. Emergency and critical care. *Vet Clin North Am Exotic Anim Pract*. 2007;10(2):345-394.

Drug Dosing in Exotic Animals: Are We Using Evidence-Based Medicine?

Marcy J. Souza, DVM, MPH, Dipl ABVP (Avian), Dipl ACVPM

Session #017

Affiliation: From the Department of Biomedical and Diagnostic Sciences, College of Veterinary Medicine, University of Tennessee, Knoxville, TN 37996, USA.

Abstract: Evidence-based medicine can be challenging when limited research has been performed to determine appropriate dosing of medications in many exotic species that are kept in captivity. This talk will discuss some inappropriate recommendations that are routinely made in clinical practice and steps to reduce the likelihood of making medical decisions based on hearsay alone.

Introduction

Ideally, dosing recommendations for all animals, including exotic animals, would be based on controlled clinical trials. Unfortunately, these types of studies are rare for exotic species. Often, we decide our dosing based on clinical experience and personal communications with other veterinarians; this strategy may or may not provide accurate information. A personal communication of success with a particular drug regimen can become ingrained in the literature despite any real evidence of success.[1] Therefore, practitioners must carefully evaluate the references for reported dosing regimens in the literature and gauge how much evidence is truly behind the recommendations.

Treatment Examples

Mycobacteriosis in pet birds

The treatment of mycobacteriosis in pet birds is controversial: some practitioners feel it appropriate to treat these animals, while others feel that euthanasia is necessary due to the potential for zoonotic infection. Regardless of whether these animals should be treated due to the zoonotic potential of *Mycobacterium* species, do practitioners have enough information to appropriately treat and cure this infection without creating drug-resistant, and potentially zoonotic, strains? And what role does owner compliance play in the treatment of these animals?

In humans, therapy for mycobacteriosis typically consists of 6 to 9 months of treatment with approximately 4 drugs given at various intervals, some daily and others 1 to 3 times per week.[2] If appropriate administration of drugs is not performed, patients quickly relapse with clinical disease and drug resistance often occurs.

In an exotic animal formulary that describes protocols used to treat mycobacteriosis in birds, 7 references are listed for drug regimens; however, none are from pharmacokinetic and pharmacodynamic studies or clinical trials.[3] Most are articles, book chapters, or proceedings, which are not necessarily peer reviewed. Additionally, in this resource, a footnote states that a clinical trial evaluating the treatment of mycobacteriosis in ring-necked doves with a multi-drug protocol was unsuccessful.[3] The lack of research behind these dosing recommendations in combination with the challenge of owner compliance certainly should factor into practitioner decisions as to whether treatment is truly best for the patient and owner.

Tramadol use in rabbits

Analgesia can be difficult to assess in animals, particularly those that are considered prey species. A reasonable amount of research has been performed examining tramadol use in dogs and cats, and recommended doses are typically 4 to 5 mg/kg PO 2 to 3 times per day. Anecdotally, a dose of 10 mg/kg PO 1-2 times per day has been recommended for rabbits. Because of this recommended dose, the author (MS) performed a research study with others to examine the pharmacokinetics of oral dosing (11 mg/kg) of tramadol in rabbits. Plasma concentrations were below those associated with analgesia in humans.[4] Another study in Hispaniolan Amazon parrots found that plasma concentrations associated with analgesia were similar to those reported in humans,[5] but we do not know the therapeutic concentrations in rabbits. Administration of 4.4 mg/kg IV led to a statistically significant reduction in the amount of isoflurane needed to maintain a surgical plane of anesthesia, but the amount was not clinically significant.[4] In a separate unpublished study, rabbits that had been spayed and treated post-operatively with tramadol (11 mg/kg PO) were noticeably in more pain than rabbits treated with tramadol (11 mg/kg PO) and meloxicam. Despite evidence that 11 mg/kg PO is unlikely to reach therapeutic concentrations, a recent review recommended a dose of 10 mg/kg PO 1 to 2 times per day based on personal experience and anecdotal reports.[6] A pain evaluation scale is currently being developed by Flecknell, and this tool could aid in the evaluation of efficacy of analgesics in rabbits. For more information about recognizing pain in laboratory animals, please visit: http://www.ncbi.nlm.nih.gov/books/NBK32656/.

Conclusion

It is up to the practitioner to be well informed on how to make decisions for patients. Critical evaluation of the literature is of utmost importance when making decisions. Although it is often easiest to perform a Web search for dosing recommendations, practitioners should make sure they know what those recommendations are based on. Are they based on a personal communication after treatment of a few patients? Or are they based on peer-reviewed research? Also, whenever given the opportunity, practitioners should contribute to clinical research. Whether it be by submitting samples, providing records review, or participating in a clinical trial, YOUR participation in the process can improve and increase the knowledge needed for the treatment of exotic pets.

References

1. Kukanich B. Clinical interpretation of pharmacokinetic and pharmacodynamic data in zoologic companion animal species. *Vet Clin North Am Exot Anim Pract*. 2011;14:1-20.

2. Centers for Disease Control and Prevention. Treatment of tuberculosis: American Thoracic Society, CDC, and Infectious Diseases Society of America. *MMWR Morb Mortal Wkly Rep*. 2003;52(RR-11). Available at: http://www.cdc.gov/mmwr/PDF/rr/rr5211.pdf. Accessed April 11, 2016.

3. Carpenter JW, Marion CJ, eds. *Carpenter's Exotic Animal Formulary*. 4th ed. St. Louis, MO: Elsevier Saunders;2013.

4. Souza MJ, Greenacre CB, Cox SK. Pharmacokinetics of orally administered tramadol in domestic rabbits (*Oryctolagus cuniculus*). *Am J Vet Res*. 2008;69:979-982.

5. Souza MJ, Sanchez-Migallon Guzman D, Paul-Murphy JR, Cox SK. Pharmacokinetics after oral and intravenous administration of a single dose of tramadol hydrochloride to Hispaniolan Amazon parrots (*Amazona ventralis*). *Am J Vet Res*. 2012;73:1142-1147.

6. Barter LS. Rabbit analgesia. *Vet Clin North Am Exot Anim Pract*. 2011:14:93-104.

How, What, and Why of Tracking the Acute-Phase Response

Carolyn Cray, PhD

Session #144

Affiliation: From the University of Miami Miller School of Medicine, Department of Pathology and Laboratory Medicine, Division of Comparative Pathology, P.O. Box 016960 (R-46), Miami, FL 33101, USA.

Abstract: Tracking the acute phase response has demonstrated clinical value in avian, reptile, and exotic mammal species and the use of acute-phase protein biomarkers is now well established in companion animals, large animals, and non-domesticated mammals. Protein electrophoresis has been a mainstay and is still a high value tool in avian species but newer methodologies allow for the opportunity to view changes in specific major acute-phase proteins. The how, what, and why of acute-phase protein testing will be discussed as well as other possible biomarkers of interest. These measures, including electrophoresis, will be contrasted to traditional blood tests.

Introduction

The acute-phase response (APR) is an integral part of the early innate immune response which is also inclusive of physical barriers, phagocytes, complement, and toll-like receptors.[1-3] The APR is a complex systematic inflammatory process that begins with a local stimulus such as a response to injury or infection and then the production of numerous cytokines which are produced in part by the first responding macrophages and monocytes. Some of these cytokines signal the liver to initiate or upregulate the expression of acute-phase proteins (APP). It has been estimated that more than 200 APP may be produced during the APR.[1] Each APP has a unique biological function from acting as an opsonin to activate complement and enhance phagocytosis (C-reactive protein, CRP), enhancing chemotaxis and aiding in tissue repair (serum amyloid A, SAA), and binding hemoglobin to minimize oxidative activity (haptoglobin, HP).[2] The goal of this response is to promote healing and restore homeostasis.[2] These proteins have been found to highly conserved with observations of these or similar proteins from mollusks to turtles to mammals.

The expression of APP can increase or decrease with an ongoing APR.[2-4] Major APP can increase 10-1000-fold and are often present in negligible levels in normal animals. The increase is rapid and often occurs within 24 hours. When a suitable response is made in the inflammatory stimuli, the decrease is often equally rapid due to a good negative feedback system and a short half-life. In many companion and large animals, SAA and CRP act as major APP. Mild and moderate APP are often present in detectable levels in normal animals. Increases may be 2-10 fold for moderate APP and less than 2-fold for minor APP. Expression is often delayed to 4-6 days after insult and elevated levels will persist after resolution of the inflammatory process. HP is often a minor to moderate APP in most species.

Albumin is classified as a negative APP as it decreases in ongoing APR. In some mammals, transferrin has also been classified as a negative APP.

APP biomarkers contrast the sensitivity of monitoring other more traditional means of measuring inflammation including the total white blood cell count, neutrophil count, and the A/G ratio. When gold standard agents of inflammation are injected into laboratory animals (e.g., turpentine oil, Freund's adjuvant, lipopolysaccharide), there can be a mild transient increase in WBC but a true leukocytosis was not observed until 4-7 days later. Serum albumin will decrease but not reaching minimum concentrations until day 5. In contrast, CRP and SAA increase several hundred fold by 48 hours. Furthermore, the features of APP – including duration of response, magnitude of response, and short half-life – make these markers very viable prognostic indicators. In addition, their application in wellness exams has also been demonstrated.

As the origin of the APP stimulus may be infection, inflammation, trauma, stress or neoplasia, APP are rarely solely diagnostic for any one disease or etiology.[2] While it may not necessarily be fair to compare these independent measures given their different timelines of expression, many studies have been conducted in dogs and horses. In a study from dogs with various inflammatory processes showed a correlation of $r = 0.44$ between WBC counts and CRP and, in a study of horses, the correlation between WBC counts and SAA was weak ($r = 0.11$).[2] In the latter study, SAA had the highest diagnostic accuracy for inflammation vs. the other measures at 75%.[2]

An ongoing APR can be reflected in fraction changes in protein electrophoresis.[1] Newer automated and semi-automated methodologies including enzyme-linked immunosorbent assay, immunoturbidity, and colorimetry has been employed to quantitate specific APP. In the antibody based protocols, methods and reagents must be individually validated for each species.[6] This must be done in a stepwise fashion with the assay of samples from known clinically normal and abnormal animals. If reactive reagents are identified and fully vetted at the bench level, the study can now include a large number of samples to determine reference intervals and assess the clinical impact of using these biomarkers with patients. Of possible future interest is the validation of lateral flow devices and other point-of-care testing for these biomarkers.

Birds

Studies in chickens have dominated the literature on APP expression in birds.[7,8] With injection of gold standard inflammatory agents, changes in SAA, transferrin, PIT-54 (analog of HP), and alpha-1 acid glycoprotein have all been documented. In our laboratory experience, finding reagents with wide avian species cross reactivity has been difficult although the colorimetric reagents used to quantitate HP appear to be valid. Increased SAA expression has been documented in falcons with aspergillosis.[9,10] It appears that this APP is a major APP in this species.

Our current impression is that protein electrophoresis provides very good sensitivity for ongoing APR. It provides a broad impression of different globulin changes and specifically quantitates albumin which is an important and sensitive negative APP in birds.[11] In a comparison of total WBC counts versus EPH abnormalities in African grey parrots (judged by a change in the A/G ratio), a marginal correlation was observed between significantly elevated WBC counts and significantly decreased A/G ratio (C. Cray, oral communication, May 2016). In 49% of the cases with a decreased A/G ratio, an elevated WBC count was found. Notably, 43% of the cases showed a normal WBC count when a significant decrease in the A/G ratio was present. Similarly, the majority of the cases showed a normal WBC count when a mild to moderate decrease in the A/G ratio was present. This data indicates a differential sensitivity between these two methods and supports the use of electrophoresis in routine clinical pathology investigations of avian species.

Reptiles

In soft-shell turtles experimentally infected with gram negative bacteria, SAA expression (at the RNA level) was found to increase 1000-fold.[12] Attempts to validate commercially available SAA antibodies in most reptiles have failed to date. Hemoglobin binding activity, quantitated by the colorimetric assay for HP, has been demonstrated in box turtles, loggerhead turtles, inland bearded dragons, and rattlesnakes (C. Cray, oral communication, May 2016).[13,14] As additional reagents for major APP become available, studies should consider the relative sensitivity of protein electrophoresis vs. APP assays in these species.

Exotic Mammals

Rats and mice have interesting differences in APP expression and mice produce another major APP named serum amyloid P.[4] There has been little to no description of APP expression in guinea pigs, hamsters, and gerbils.

CRP expression has been described in rabbits with suspected *Encephalitozoon cuniculi* infection.[15] Elevated CRP levels acted as an adjunct diagnostic test to improve the specificity of serologic titers against this organism and also served as a prognostic indicator.[16] Recent studies indicate also the utility of newly described SAA-1 reagents (C. Cray, oral communication, May 2016). In a study inclusive of samples from various clinically abnormal rabbits (C. Cray, oral communication, May 2016), the sensitivity and specificity of CRP for clinical abnormality was 92% and 93%, respectively. The correlation between SAA-1 and CRP was significant, $r = 0.72$, $P < 0.0001$. While WBC counts were found to be specific (100%), the sensitivity was only 25%.

Ferret SAA was cloned and found to be similar to other mammalian SAA. In studies from our laboratory, SAA was found to be increased 2-5 fold in clinically abnormal ferrets.[17] A 10-fold increase in HP was observed in a ferret with myeloma.

Other Species

APP expression has been well documented at the RNA level in several species of fish. Attempts to document expression in a trauma model in koi were not successful and may have been related to lack of reagent cross reactivity and experimental design.[18] Reference intervals have been generated for the white-spotted bamboo shark and some commercial CRP reagents have been validated for use in the bonnethead shark (C. Cray, oral communication, May 2016).[19] A commercial ELISA for SAA quantitation has also been validated for use in salmon. A wide range of published and ongoing studies have validated the use of APP in many wildlife species and in some species, these tests have been implemented as part of the standard of care for captive animals, for prognostic value in rehabilitation setting, and in population studies of free-ranging animals.

Other Markers

The field of proteomics is rapidly growing so it is expected many more biomarkers will be identified to gauge the acute-phase response and assist in the diagnosis of specific diseases. In a spectroscopy study of samples from falcons with aspergillosis, 3-hydroxybutytrate and LDL are significantly altered versus normal falcons.[20] Lipoprotein fractions were also found to be significantly altered in clinically abnormal cownose rays.[21] These observations underlie the opportunities for further study in these areas in other species.

References

1. Kaneko JJ. Serum proteins and the dysproteinemias. In: Kaneko JJ, Harvey JW, Bruss ML, eds. *Clinical Biochemisty of Domestic Animals*. 5th ed. San Diego, CA: Academic Press. 1997;117-138.

2. Cray C. Acute phase proteins in animals. *Prog Mol Biol Transl Sci.* 2011;105:113-150.

3. Eckersall PD. Proteins, proteomics, and the dysproteinemias. In: Kaneko JJ, Harvey JW, Bruss ML, eds. *Clinical Biochemistry of Domestic Animals*. 6th ed. San Diego, CA: Academic Press;2008:117-155.

4. Cray C, Zaias J, Altman NH. Acute phase response in animals: a review. *Comp Med.* 2009;59:517-526.

5. Belgrave R, Dickey M, Arheart KL, et al. Assessment of serum amyloid A (SAA) testing and its clinical application in a specialized equine practice. *J Am Vet Med Assoc*. 2013;243:113-119.

6. Cray C. Biomarkers of inflammation in exotic pets. *J Exot Pet Med.* 2013;22:245-250.

7. O'Reilly EL, Eckersall PD. Acute phase proteins: a review of their function, behaviour, and measurements in chickens. *World Poult Sci J.* 2014;70:27-44.

8. Chamanza R, van Veen L, Tivapasi MT, et al. Acute phase proteins in the domestic fowl. *World Poult Sci J.* 1999;55:61-71.

9. Caliendo V, McKinney P, Bailey T, et al. Serum amyloid A as an indicator of health status in falcons. *J Avian Med Surg.* 2013;27:83-89.

10. Fischer D, Van Waeyenberghe L, Cray C, et al. Comparison of diagnostic tools for the detection of aspergillosis in blood samples of experimentally infected falcons. *Avian Dis.* 2014;58:587-598.

11. Cray C, Wack A, Arheart KL. Invalidty of albumin measurement of specimens from clinically ill birds by bromcresol green methodology. *J Avian Med Surg.* 2011;25:14-22.

12. Zhou X, Wang I, Feng H, et al. Acute phase response in Chinese soft-shelled turtle (*Trionyx sinensi*) with *Aeromonas hydrophila* infection. *Dev Comp Immunol.* 2011;35:441-451.

13. Flower JE, Byrd J, Cray C, Allender M. Plasma electrophoretic profiles and hemoglobin binding protein reference intervals in the Eastern box turtle (*Terrapene carolina carolina*) and influences of age, sex, season, and geography. *J Zoo Wildl Med.* 2014;45:836-842.

14. Dickey M, Cray C, Norton T, et al. Assessment of hemoglobin binding protein in loggerhead sea turtles (*Caretta caretta*) undergoing rehabilitation. *J Zoo Wildl Med.* 2014;45:700-703.

15. Cray C, Rodriguez M, Fernandez Y. Acute phase protein levels in rabbits with suspected *Encephalitozoon cuniculi* infection. *J Exot Pet Med.* 2013;22:280-286.

16. Cray C, McKenny S, Perritt E, Arheart KL. Utility of IgM titers with IgG and CRP quantitation in the diagnosis of suspected *Encephalitozoon cuniculi* infection in rabbits. *J Exot Pet Med.* 2015;24:356-360.

17. Ravich M, Johnson-Delaney C, Kelleher S, et al. Quantitation of acute phase proteins and protein electrophoresis in ferrets. *J Exot Pet Med.* 2015; http://dx.doi.org/10.1053/j.jepm.2015.04.010.

18. Christiansen EF, Cray C, Lewbart G, Harms C. Plasma protein electrophoresis and acute phase proteins in koi (*Cyprinus carpio*) following exploratory celiotomy. *J Exot Pet Med.* 2015;24:76-83.

19. Krol L, Allender M, Cray C. Plasma proteins and selected acute phase proteins in white-spotted bamboo sharks (*Chiloscyllium plagiosum*). *J Zoo Wildl Med.* 2014;45:782-786.

20. Pappalardo L, Hoijemberg PA, Pelczer I, et al. NMR-Metabolomics study on falcons affected by aspergillosis. *Curr Metabolomics.* 2016;2:155-161.

21. Cray C, Rodriguez M, Field C, et al. Protein and colesterol electrophoresis of plasma samples from captive cownose ray (*Rhinoptera bonasus*). *J Vet Diagn Invest.* 2015;27:688-695.

Section 33

Wednesday Hybrid Track

Louisa Asseo, DVM and
James Brogan, DVM, Dipl ABVP, CertAqV
Moderators

Development of an Online Tumor Database for Zoological and Exotic Species: Roundtable Discussion

Ashley Zehnder, DVM, Dipl ABVP (Avian), PhD,
Tara M. Harrison, DVM, MPVM, Dipl AVZM, Dipl ACVPM,
Cassondra Bauer, DVM, MS, Dipl ACVPM,
Ryan Colburn, DVM,
Joanne Paul-Murphy DVM, Dipl ACZM, Dipl ACAW,
Michelle G. Hawkins, VMD, Dipl ABVP (Avian),
Catherine Pfent, DVM, PhD

Session #048

Affiliation: From Stanford University, School of Medicine, Stanford, CA 94305, USA (Zehnder), North Carolina State University, College of Veterinary Medicine, Raleigh, NC 27607, USA (Harrison), WIL Research, Ashland, OH 44805, USA (Bauer), John Ball Zoo, Grand Rapids, MI 49504, USA (Colburn), School of Veterinary Medicine, University of California, Davis, CA 95616, USA (Paul-Murphy, Hawkins), Lincoln Memorial University-College of Veterinary Medicine, Harrogate, TN 37752, USA (Pfent).

Diagnoses of cancer are increasing in frequency in exotic animal medicine and zoological institutions. Increasingly, there are parallels between human and animal cancer and therapeutics.[1,2] Further, improved preventive medicine and earlier diagnoses of cancer in zoological species has expanded the therapeutic options in zoo animals. Cancers have been diagnosed and treated in fish, reptiles, birds, and all forms of mammals.[3-6,7] Reported cancers include lymphoma, papillomas, carcinomas, sarcomas, and melanomas. There are likely to be many more cases that have gone unreported, whether as a case report or case series. Treatments in zoo animals have been facilitated by use of vascular access ports for delivery of chemotherapy, and for anesthesia for radiation treatments.[8-10] Through operant conditioning, and the development of newer, safer medications, treatment of zoological species is increasingly feasible. The Exotic Species Cancer Research Alliance, in collaboration with Stanford University, North Carolina State University, and U.C. Davis is developing a data base for clinicians in private practice and in zoological institutions to contribute cases, query for parameters related to tumor type, treatment types, adverse effects of treatment and outcomes. Institutions are encouraged to participate through case contributions to this database, which will include diagnoses, plus annotation of any therapeutics employed and outcomes observed. Through these collaborative efforts, we hope to discover trends in cancer incidence, better understand treatment and therapeutic options for exotic and zoological species, and provide a long-term archive of study material for future research.

References

1. Fukumoto, S, Hanazono K, Fu DR, Endo Y, Kadosawa T, Iwano H, Uchide T. A new treatment for human malignant melanoma targeting L-type amino acid transporter 1 (LAT1): a pilot study in a canine model. *Biochem Biophys Res Commun*. 2013;439:103-108.

2. Ranieri G, Gadaleta CD, Patruno R, Zizzo N, Daidone MG, Hansson MG, Paradiso A, Ribatti D. A model of study for human cancer: spontaneous occurring tumors in dogs. Biological features and translation for new anticancer therapies. *Crit Rev Oncol Hematol*. 2013;88:187-197.

3. Abou-Madi N, Kern TJ. Squamous cell carcinoma associated with a periorbital mass in a veiled chameleon (*Chamaeleo calyptratus*). *Vet Ophthalmol*. 2002;5:217-220.

4. Anderson ET, Kennedy-Stoskopf S, Sandy JR, Dorn B, Boyette T, Harms CA. Squamous cell carcinoma with vascular invasion in a diamondback rattlesnake (*Crotalus adamanteus*). *J Zoo Wildl Med*. 2010;41:745-748.

5. Bertelsen MF, Steele SL, Grondahl C, Baandrup U. Pheochromocytoma in a white rhinoceros (*Ceratotherium simum*). *J Zoo Wildl Med*. 2011;42:521-523.

6. Dillberger JE, Citino SB, Altman NH. Four cases of neoplasia in captive wild birds. *Avian Dis*. 1987;31:206-213.

7. Lombardini ED, Hard GC, Harshbarger JC. Neoplasms of the urinary tract in fish. *Vet Pathol*. 2014;51(5):1000-1012.

8. Kent MS. The use of chemotherapy in exotic animals. *Vet Clin North Am Exot Anim Pract*. 2004;7:807-820.

9. Rassnick KM, Gould WJ, Flanders JA. Use of a vascular access system for administration of chemotherapeutic agents to a ferret with lymphoma. *J Am Vet Med Assoc*. 1995;206:500-504.

10. Roberts WG, Klein MK, Loomis M, Weldy S, Berns MW. Photodynamic therapy of spontaneous cancers in felines, canines, and snakes with chloro-aluminum sulfonated phthalocyanine. *J Natl Cancer Inst*. 1991;3:18-23.

Characterization of Gout Versus Pseudogout Crystals in Exotic Animals

Kevin Keel, DVM, MS, PhD, Dipl ACVP, and
Ash Sundaram, MS

Session #075

Affiliation: From the UC Davis School of Veterinary Medicine, Davis, CA 95616, USA.

Abstract: Gout, pseudogout and hydroxyapatite deposition disease (HADD; syn tumoral calcinosis, calcinosis circumscripta) are characterized by the accumulation of crystals in soft tissues and may represent a diagnostic challenge in the evaluation of histopathology material. These lesions occur in a variety of species and can be difficult to differentiate in some instances. Gout is defined by the deposition of monosodium urate crystals whereas pseudogout and HADD are typically associated with accumulation of calcium pyrophosphate dihydrate ($Ca_2P2O_7.2H2O$) or hydroxyapatite ($Ca_{10}(PO_4)_6(OH)_2$), respectively. The appearance of pseudogout has been variably described as amorphous, globular to fibrillar material that is basophilic to amphophilic with giant cells or chondroid metaplasia. Descriptions of gout are just as varied and range from amorphous, eosinophilic material to feathery, amphophilic material. As with the deposition of calcium salts, urate deposition often incites a foreign body inflammatory response. We hypothesized that the microscopic appearance of gout, pseudogout and HADD on HE-stained sections is insufficient to differentiate these lesions with a significant level of specificity. To test this hypothesis we used Von Kossa stain and nonaqueous alcoholic eosin stain (NAES) to evaluate cases diagnosed with gout (n=14), pseudogout (n=7) or both (n=1). Thirteen birds and nine reptiles were represented. The NAES method prevented the dissolution of urates that often occurs in typical histologic processing. The Von Kossa stain allowed the confirmation of divalent cations, represented by calcium salts in pseudogout.

References

1. Shidham V, Chivukula M, Basir Z, Shidham G. Evaluation of crystals in formalin-fixed, paraffin-embedded tissue sections for the differential diagnosis of pseudogout, gout, and tumoral calcinosis. *Mod Pathol.* 2001;14:806-810.

2. Jones YL, Fitzgerald SD. Articular gout and suspected pseudogout in a basilisk lizard (*Basilicus plumifrons*). *J Zoo Wildl Med.* 2009;40(3):576-578.

Practical Application of Photobiomodulation (Laser Therapy) in Exotic Pet Practice

Robert D. Ness, DVM

Session #333

Affiliation: From Ness Exotic Wellness Center, 1007 Maple Ave, Lisle, IL 60532, USA.

Laser therapy is becoming a common modality in veterinary practice. Therapeutic lasers are being routinely used for the treatment of everything from simple surgical incisions to severe osteoarthritis. Potential applications continue to expand as practitioners utilize this revolutionary technology to treat virtually any condition that exhibits pain or inflammation. In addition, many practices routinely incorporate laser therapy in any wound management or arthritis protocol. It would not be surprising to find this technology to be considered the standard of care in veterinary practice in the not so distant future.

Photobiomodulation, also known as laser therapy, utilizes a specific wavelength of near-infrared light to penetrate tissues and elicit a biochemical and biological response. The laser light differs from ordinary visible light by being monochromatic (single wavelength), collimated (tight narrow beam) and coherent (uniform orderly beam). As a result the laser is an amplifier of light which sustains a stimulated emission of photons (energy packets of light). The resultant effects of photobiomodulation include reduced pain, reduced inflammation and accelerated healing. Effective treatment depends on delivering an appropriate dose of laser energy into the affected tissue. Patients with acute inflammatory or traumatic conditions typically respond with a single or few treatments sessions at relatively low target doses. Chronic or deep tissue conditions require a higher target dose and more frequent treatment sessions, with some necessitating ongoing maintenance therapy.

Protocol settings for exotic species were typically extrapolated from existing protocols for cats and dogs, but some laser companies have established protocols and incorporated the settings for bird and exotic pets into their therapeutic laser units. Consideration for the anatomic and physiologic differences in these exotic species required further adjustments of the protocols. For example, adjustments are made to further decrease the avian settings to take into account their thin, delicate skin and feather architecture, as well as their higher metabolic rates. Conversely, the dosage settings for reptiles are relatively increased to accommodate for their thicker skin and lower metabolic rates. The small mammals required fewer modifications except for body size and weight when utilizing the feline settings. As a precaution, certain Laser units for exotics have a pulsed frequency and generally lower power pre-sets than the comparable protocols for traditional companion species.

Success in treating conditions depends on delivering an appropriate dose of laser energy into the affected tissue. Insufficient dosing results in treatment failure. Overdosing wastes time and practice resources, however it is not detrimental to the patient. Target doses are expressed in J/cm^2. One joule (J) is the energy delivered with one Watt of power in one second. Established target doses are 3-4 J/cm^2 for superficial conditions (conditions in tissue that we can see) and 8-10 J/cm^2 in deep conditions (conditions beneath the skin). Power and time are inversely proportional in calculating the total energy (J) delivered. Treatment time can be reduced by increasing power. Likewise, treatment time increases if the power (Watts) is reduced. When treating superficial conditions with a lower target dose, power usually remains relatively low since achieving a target dose of 3 - 4 J/cm^2 does not take long. Many superficial conditions affect sensitive tissue and a lower power setting is more comfortable for the patient. When treating deep tissue conditions, power is increased to reduce the time required to achieve a target dose of 8-10 J/cm^2. Some laser treatment protocols, such as wound or abdominal disorders, require an estimation of the treatment area when calculating the target dose of a laser treatment. A simple technique for estimating the patient's surface treatment area is by comparing the size of the area to be treated relative to a 3 × 5" index card, which has an approximate area of 100 cm^2.

Small Exotic Mammal Laser Therapy

Exotic small mammal protocols are modified versions of the established feline settings for many of the common conditions seen in both types of patients. Similar to other companion animals, therapeutic dosages established for trauma and wound settings are based on lesion size, typically starting at 1-2 J/cm^2. These settings are used for pododermatitis in rabbits and traumatic wounds in ferrets, for example. However, photobiomodulation target dosing protocols for conditions such as arthritis, pain, and swelling/edema are based on patient size and anatomic location of the body. Spondylosis in aged rabbits resulting in hind limb weakness and paresis responds well to arthritic dosage settings, at a therapeutic dose up to 10 J/cm^2. The quality of life in these older patients is greatly improved when utilizing the laser at the arthritis and chronic pain setting by reducing pain and inflammation, which results in improved mobility. This is also the setting used for the unfortunate condition of spinal or pelvic trauma experienced in rabbits, however these cases may require more frequent laser sessions for a positive outcome. Photobiomodulation is even beneficial when incorporated in the treatment protocol for gastrointestinal stasis in rabbits and guinea pigs. A therapeutic target dose of 7-10 J/cm^2 is recommended, with the area based on treating the entire abdominal region. Utilizing photobiomodulation in these cases has greatly improved the recovery of these patients in practice by reducing inflammation and providing abdominal pain relief. Similar results are observed in cases of cystic calculi in ferrets, rabbits and guinea pigs.

Avian Laser Therapy

Avian patients equally benefit from the use of photobiomodulation. More precautions should be taken when treating a bird then with most other species because of their thin dermis and feathering. As a result, the established settings in birds are relatively lower to comparable indications in mammal species. Wound and trauma settings are quite useful in treating pododermatitis in pet birds and raptors in addition to self-mutilated wounds commonly encountered in Quaker parrots and cockatoos. These superficial lesion cases require a target dosage of 2-4 J/cm^2 for treatment, with the number of sessions and frequency dependent on the severity and chronicity of the lesion. Of course, the underlying conditions must also be addressed, but the laser treatments will expedite the recovery. In fact, laser settings have been provided to facilitate recovery from feather picking by stimulating the quiescent feather follicles and reduce inflammation in those cases involving folliculitis. Feather follicle stimulation target dosages can range from 2-3 J/cm^2 to as high as 10 J/cm^2, depending on severity and chronicity. Protocols for fractures, arthritis and other painful conditions are based on the size of the bird as well as the anatomic location of the problem, but typically range between 4 – 8 J/cm^2 in most cases. In clinical experience, fractures which are treated with photobiomodulation tend to recover approximately 25-50% faster than those managed by traditional means alone.

Reptile Laser Therapy

Reptile protocols are approached differently at this time, due to the variety of species seen in practice and the wide range of size and shapes. Individual consideration must be implemented for variables such as the shell of a tortoise, thick scales of a blue-tongue skink or the thin dermis of a leopard gecko lizard. However, a wide variety of conditions can be readily treated in any of these reptile species. Cases that are generally classified as superficial lesions in terms of laser protocol settings include rat bite wounds, blister disease, thermal burns, and infectious stomatitis. These lesions can be treated with an initial target dose of 1-3 J/cm^2 in uncomplicated cases to as high as 8-10 J/cm^2 for rat bite lesions having deep wounds with significant muscular involvement. Shell fractures in turtles and tortoises heal more rapidly with use of photobiomodulation, starting with a target dose of 4-6 J/cm^2 for cracked shell to 10-12 J/cm^2 for deep shell wounds penetrating the coelomic cavity. Deep seated conditions, such as metabolic bone disease, fractures and arthritis, would be treated with a target dose ranging from 6-10 J/cm^2, or even higher if severe and chronic. As with birds and mammals, anything that causes pain or inflammation can be addressed with laser therapy.

Photobiomodulation is indicated in many conditions commonly encountered in avian and exotic practice. As with our dog and cat patients, the incorporation of a therapeutic laser in daily practice will greatly enhance the quality of life in so many patients by expediting the healing of lesions as well as reducing inflammation and managing pain.

Establishing a Fish Practice in an Existing Exotics Practice

Mike Corcoran, DVM, CertAqV, and
Katie Haman, DVM, MSc

Session #218

Affiliation: From VCA Wakefield Animal Hospital, 19 Main St., Wakefield, MA 01880, USA and Washington State Department of Fish and Wildlife, 1111 Washington Street NW, Olympia, WA 98501, USA.

Abstract: Adding aquatic animals to an existing exotic animal practice is relatively simple and can add a fun, challenging aspect to the practice. This article will share the authors' experiences and give some advice to practitioners wanting to explore this field.

Introduction

Exotic animal practitioners are experienced with taking what is known regarding more commonly treated species and applying the principles of good quality medicine to less frequently studied species. This often requires learning new anatomy, physiology and husbandry. Fish and other aquatic animals are no different. Continuing education is available from a number of different sources. Aside from a small investment in new equipment and some marketing, the addition of aquatic animal clients to an existing exotic animal practice is seamless.

Supplies Required to Add Aquatic Animal Patients

Very basic supplies required to start seeing fish as patients requires a minimal investment. Of course, as the practice is expanded more specialized supplies may be beneficial or required for additional procedures. In this section we will discuss materials in some depth and provide a table listing basic minimal supplies and additional supplies that are useful as you come into contact with more advanced cases and expand the practice. We will also discuss the role of veterinarians in aquaculture and fish population health. Expanding a practice to include not only pet fish, but also food fish is easily accomplished; the most difficult component will most likely not be the medical aspect, but instead mastering the elements of husbandry and fish culture that are frequently the primary cause of clinical disease, both in pet and food fish.

Minimum supply list

Many of the items on this list will already be available, but it's wise for some to be dedicated and it's also included on this list to highlight some important considerations. Gloves, cover slips and suture scissors will already be in a practice, but there are some special considerations. Gloves should always be used when handling aquatic animals (for the protection of the patient and the staff). Gloves used for handling fish should not contain powders or lotions. These can adversely affect the protective slime coat of the patient. Nitrile gloves are a great choice. If powdered gloves are the only glove available, then the powder must be rinsed before contacting the fish. Plastic cover slips are recommended for obtaining skin (slime coat) cytology samples due to the decreased risk of iatrogenic trauma during the procedure. Suture scissors or iris scissors are used to obtain gill biopsy samples as part of the basic examination. It is advisable to have a dedicated pair (or several) that can be decontaminated appropriately and will not be subject to use with adhesive tapes or other non-clinical use.

Other supplies required to get started include some type of aquarium or holding tank for sedation, air stones, nets, water testing kits and some specific medications. These will be discussed in further detail in the following sections.

Holding tanks and aquariums

There is no telling how an aquatic animal will be presented to you at the hospital. Fish have been brought for appointments in the normal bowl or aquarium that acts as their habitat and have also been brought in plastic bags, coolers, coffee mugs and anything else that will hold water. The practitioner should be prepared to receive animals in many different ways and be able to hold them in the hospital safely for many hours. In a beginning practice, many inexpensive options exist. Try to have varying sizes for different fish from bettas to koi. Small "Critter Keepers" can be used for smaller fish and any type of plastic storage container can be utilized for larger fish. Some type of cover should be available to prevent escape during the excitatory phase of sedation and the container should allow easy handing and good visualization of the patient.

As the practice develops, it may also be wise to invest in more semi-permanent solutions to allow for longer stays on surgery patients or even hospitalization of animals. Aquariums varying from 20 gallons to 40 gallons could be kept cycled at all times in a designated treatment room if the case load is large, or collapsible containers are available. Filtration will be discussed later, but should be considered when selecting larger and more permanent setups. For large fish, collapsible show tanks for koi are relatively inexpensive and take up little storage space even for several hundred gallons. Smaller store display tans or breeder tanks can be used as well providing they allow for disinfection between animals.

If dealing with high value koi, there are some specialized pieces of equipment that should be considered. Keepers of these fish use certain tools to ensure gentle handling of show quality fish, and your use of the same helps distinguish you as a professional in their eyes. Round holding tanks are standard for handling koi out of the pond. The use of sock nets for transportation between tanks allows decreased stress and decreased chance of trauma. These items can be obtained from many breeders or pond suppliers. Using the owner's own equipment should always be the first choice for sterility, but the practitioner working with these animal frequently should have their own supplies when needed.

Air stones and filtration

Maintaining good water quality is an essential part of aquatic animal health. Water quality testing will be discussed below, but some consideration must be given to how to maintain good quality. In a new practice that is seeing only occasional outpatients, a simple air stone system can provide aeration in the holding tank or in the owner's shipping container while the patient is in the hospital for the appointment. Good air stones are available for as cheap as $20 from any pet store. Battery operated models are available for transportation.

For longer duration stays, some type of mechanical, biological and/or chemical filtration system may be required. An understanding of filtration systems and the needs of the practice will dictate the type of system required and how to install the system.

Medications

Many medications utilized in the practice will be useful for aquatic animals. Formularies have good doses for Meloxicam, Ceftazidime and Enrofloxacin among many others. Referencing formularies is wise to utilize what is already in the inventory.

For food fish, FDA regulations on the use of drugs/chemicals in aquaculture must be considered. There are only a few drugs approved for use in aquaculture (http://www.fda.gov/AnimalVeterinary/DevelopmentApprovalProcess/Aquaculture/ucm132954.htm). The use of prescriptions and veterinary feed directives (VFDs) is commonplace in aquaculture, and with the impending changes to VFDs (2017), aquaculture facilities will need to rely more on veterinary services to obtain medications and treat sick fish populations.

In the private practice, there are some additional drugs that will need to be ordered specifically for fish. Most contact with fish will require sedation for a thorough examination and any required treatment. MS-222 (http://www.wchemical.com/tricaine-s-ms-222.html) is a very safe anesthetic agent that can be used in the water to sedate most fish. It has a relatively short established withdrawal time (21 days) for food fish and a wide margin of safety in most animals. It is inexpensive and has a good shelf life. With most freshwater sources it must be buffered during use to counteract the acidity. Sodium bicarbonate in the form of ordinary baking soda should be purchased at the same time. If used in saltwater, MS222 need not be buffered. Always test the pH of the MS222 solution before sedating or anesthetizing fish.

Another anesthetic that is commonly used in aquaculture is Aqui-S 20E (http://www.aqui-s.com/index.php/aqui-s-20e). Currently available under an INDA, Aqui-S 20E can be used in food fish in the field for immediate release, or in an aquaculture setting with 72-hour withdrawal period.

Some type of water conditioner should also be on hand at all times. Read the bottle to make sure that it will remove chloramines in addition to chlorine. Always encourage owners to bring enough extra water from home to do any testing and use the same source water for anesthesia, but be prepared to utilize dechlorinated water from the hospital when this doesn't happen.

External parasites are a common finding in fish medicine. There are a number of water based medications that can be used for the parasites. Western Chemical (http://www.wchemical.com) and National Fish Pharmaceuticals (http://www.nationalfishpharm.com/) are good sources. Again, the specific caseload and regional diseases will dictate what products are most useful.

Water testing supplies

Of course husbandry is important to address with any patients and improper husbandry is at least a contributing factor to a great deal of exotic animal disease. Good water quality is vital to the health of aquatic animals. A good quality water testing kit should be on hand. At a minimum the aquatic animal veterinarian should be able to evaluate temperature, pH, hardness, alkalinity and the biological filter (ammonia, nitrite and nitrate) using an accurate kit. As the practice expands, other parameters may become important. Later tests could include dissolved oxygen, various minerals and electrolytes, conductivity and oxidation/reduction potential.

Kits come in many forms. As a professional, the accuracy is important. An inexpensive but fairly accurate option would be a kit like the API Freshwater Master Kit (pH, ammonia, nitrite and nitrate) and individual tests for alkalinity and total hardness (Available through API, Red Sea and other companies). As soon as possible it's wise to upgrade from these kits to get a more professional kit. HACH (http://www.hach.com) makes a middle range professional kit that costs just over $300, the FF-1A that is recommended by the author (MC).

Husbandry and water quality is perhaps one of the most important aspects of fish health, especially in a captive rearing environment like a hatchery or aquaculture facility. In aquaculture, water quality doesn't just include above mentioned variables that are easily measured, but also factors such as flow, turbidity, pounds of fish per gallons/minute, and density index (or number of fish relative to rearing space and flow).

Most often, diseases seen in a population of fish are due to suboptimal environmental conditions that lead to increased stress and decreased immune function, making the population more susceptible to pathogens that otherwise do not cause disease. There are numerous examples of this in aquaculture, and changes to husbandry and/or fish culture can drastically improve fish health and minimize the need to medically treat.

Including Aquaculture in Your Practice

To include an aquaculture or hatchery facility as a client in your practice you will need to visit the farm. Making a mobile fish practice is simple – a kit including slides, coverslips, scalpel blades, an anesthetic, gloves, stain (Diff Quick), and a microscope should be sufficient for most visits to an aquaculture facility. This enables the veterinarian to collect the basic samples (skin scrapes, gill clips) and conduct necropsies to look for pathogens in internal organs. Further diagnostics, if warranted, can be done back in the clinic or by sending samples to appropriate laboratories.

Working as a veterinarian in aquaculture requires more than medical knowledge and diagnostic capabilities. One must also appreciate the complexities of fish culture, husbandry, feeding, and the life history of the species being reared. Many diseases are preventable, and if conditions are addressed and changed accordingly, will not manifest as population-level disease and will not require medical treatment. In this capacity, it is the role of the veterinarian to work closely with fish culturists to ensure best practices are employed to maximize fish health.

The use of drugs in aquaculture and food fish is regulated by the FDA. As a veterinarian working in aquaculture, you must know what drugs are approved for what species and diseases and employ them accordingly. Withdrawal periods must always be met. There are few drugs approved for use in aquaculture, so options are limited. This further highlights the importance of preventing disease and maintaining good population health rather than relying on medications.

Expanding your practice to include aquaculture can be very rewarding and lucrative. Aquaculture is a rapidly growing market, and can sustainably provide food for a growing human population. Veterinarians can play a vital role in this, helping ensure healthy fish and sustainable practices as the field expands.

Marketing a Fish Practice

Start marketing your fish practice to the existing clients in your practice. Fish are the most common pet in the United Sates, there's a good chance that many of your existing clients have fish in addition to the pets you treat. You already have a relationship and trust established. Use your existing social media to feature any wet pets or share interesting articles regarding aquatic animal medicine.

Contact pet stores, pond stores and aquarium stores in the area, starting with any with which you have an existing relationship. Most fish owners try to utilize the pet stores first when dealing with an illness, not always because it is cheaper, but often because they are unaware of veterinarians as an option. Many store employees welcome the chance to refer to a veterinarian and report that they are uncomfortable recommending the OTC products in the store. In addition to a referral source, offering a good discount on services to the store can give you additional opportunities to learn while demonstrating to the staff of the stores what can be offered.

It's also a good idea to research hobbyist groups in the area. Koi clubs are very common as are aquarium hobbyist groups. Go to meetings when possible and offer to be a speaker. Not only will they feel more comfortable contacting you with any concerns with their own fish, but these groups are often contacted by outsiders for advice regarding fish disease. Send letters or other announcements to other veterinarians in the area for referrals and see about listing on local veterinary medical associations.

If interested in the ever-growing aquaculture field and population health of fish, reach out to local facilities. If you are unaware of aquaculture facilities or hatcheries in your area, contact the state department of fish and wildlife or natural resources for a list of licensed facilities. A local or regional US Fish and Wildlife Service office will also have information on federal facilities and potentially private aquaculture businesses in the area. With the pending changes to the FDAs rules on VFDs, there is going to be an increased need of veterinary services in aquaculture.

Finally, there are groups (listed below) that are good sources for support and education.[1-5] They also have listings of veterinarians under a find-a-vet section of their websites.

References

1. American Association of Fish Veterinarians. Available at: http://www.fishvets.org. Accessed July 6, 2016.
2. World Aquatic Veterinary Medical Association. Available at: http://www.wavma.org. Accessed July 6, 2016.
3. Roberts HE. *Fundamentals of Ornamental Fish Health*. Ames, IA: Wiley-Blackwell;2010.
4. Noga EJ. *Fish Disease: Diagnosis and Treatment*. Ames, IA: Wiley-Blackwell;2010.
5. Saint-Erne N. *Advanced Koi Care for Veterinarians and Professional Koi Keepers*. Glendale, AZ: Erne Enterprises;2010.

Section 34

Hybrid Posters

European and American Clinical Cooperatives: A Novel Veterinary Student Study Abroad Program

Steve Thompson, DVM, Dipl ABVP (Canine and Feline),
Zdenek Knotek, MVDr, PhD, Dipl ECZM (Herpetology),
Vladimir Jekl, MVDr, PhD, Dipl ECZM (Small Mammal)

Session #034

Affiliation: From Departments of Veterinary Clinical Sciences and Veterinary Teaching Hospital, College of Veterinary Medicine, Purdue University, 625 Harrison St, West Lafayette, IN 47907, USA (Thompson), and the Avian and Exotic Animal Clinic. Faculty of Veterinary Medicine, University of Veterinary and Pharmaceutical Sciences, Palackeho trida 1946/1, Brno, Czech Republic 612 42, CZ (Knotek, Jekl).

Purdue University's College of Veterinary Medicine (PVM) and the University of Veterinary and Pharmaceutical Sciences Brno (UVPS) in the Czech Republic entered into a sister school relationship with a memorandum of understanding in 2008. The focus was initially to develop clinical cooperatives with visitor exchanges hosting faculty and veterinary students. Initially, several fourth- and fifth-year UVPS Brno veterinary and food hygiene students visited Purdue through their Erasmus charter for 1 month each year and rotations were included through PVM and the College of Agriculture.

The PVM's inaugural study abroad course for credit expanded on this positive cooperation and became a 3-week faculty-led program with a focus on exotic pet medicine and surgery as well as exposure and participation in zoological and conservation medicine. Twenty-nine veterinary students who had completed their first or second year, have participated in this 3-week, 3-credit course. UVPS serves as the home base for many networked cooperatives that have expanded during visits in 2010, 2011, 2013 and 2015. The heart of the program remains 10-12 days of exotic pet clinics with reptiles, small mammals and birds at UVPS with an annual caseload exceeding 5000 visits. Students were responsible for patient care of outpatient and hospitalized pets and observed numerous surgical procedures. All cases were discussed in the clinic and in-depth after hours. In 2015, Prof Knotek and Dr. Jekl became adjunct professors within the Department of Veterinary Clinical Sciences in the PVM in recognition of their ongoing teaching cooperatives with PVM students. On-site experiences also have occurred with wildlife and care of the collection at Zoo Brno, a large metropolitan zoo with an adjacent wildlife rehabilitation facility and Vienna's Tiergarten Schonbrunn, the oldest zoo in the world. Behind-the-scenes tour opportunities have developed with Zoo Praha as well. Networking has expanded to include the equine and camelid services from UVPS in Brno. Prof Knotek's joint appointment with the University of Veterinary Medicine in Vienna (VetMedUni), the second oldest veterinary university in the world, allowed students to work with avian species, reptiles and aquaculture species in addition to an active spring wildlife rehabilitation service on site. Networking in Austria has since expanded to include the Messerli Research Institute and their behavior studies through the Clever Dog, Corvid and Kea labs. It is anticipated that the program will include VetMedUni rotations with exotic small mammals, equines and imaging in 2017.

UVPS has concurrent veterinary masters and doctoral degree study programs taught in English so communication with faculty has been very good considering the native language is not English and none of the PVM students or faculty have spoken Czech. The International Veterinary Student Association (IVSA) Vienna chapter has been helpful in expanding social interactions with veterinary students in Austria. PVM added a formal IVSA chapter after initial cooperation with IVSA Vienna. Cultural opportunities have included the Gregor Mendel museum in Brno, the birthplace of the study of genetics, as well as morning training of the Lipizzaner stallions at the Spanish Riding School in Vienna.

Acknowledgments: The authors would like to acknowledge the support of their rectors, deans, department heads, international program offices, colleagues and spouses. Purdue University Study Abroad and International Learning grants, Purdue University Study Abroad Initiative scholarships, Indiana VMA, SAVMA Study Abroad scholarships, PVM Spates Pet Bird Endowment and PVM Office of Lifelong Learning have provided funding and assistance that has offset some of the faculty and student costs for this study abroad course while also enhancing UVPS faculty visits to benefit PVM students onsite in West Lafayette. Brno faculty have been speakers related to exotic pets and diversity during Purdue CE conferences and national SAVMA and APVMA symposia.

AUTHOR INDEX

AUTHOR INDEX